D1561489

SIGILL : COLL : CARLETON
DECLARATIO SERMONUM TUORUM ILLUMINAT
NORTHFIELD, MINN. A.D. 1866

LIBRARY

Gift of

The Bush Foundation

WITHDRAWN

Crisis and Knowledge

Crisis and Knowledge

The Upanishadic Experience and Storytelling

YOHANAN GRINSHPON

OXFORD
UNIVERSITY PRESS

OXFORD
UNIVERSITY PRESS

YMCA Library Building, Jai Singh Road, New Delhi 110 001

Oxford University Press is a department of the University of Oxford.
It furthers the University's objective of excellence in research, scholarship,
and education by publishing worldwide in

Oxford New York

Auckland Bangkok Buenos Aires Cape Town Chennai
Dar es Salaam Delhi Hong Kong Istanbul Karachi Kolkata
Kuala Lumpur Madrid Melbourne Mexico City Mumbai Nairobi
São Paulo Shanghai Taipei Tokyo Toronto

Oxford is a registered trademark of Oxford University Press
in the UK and in certain other countries

Published in India
by Oxford University Press, New Delhi

© Oxford University Press 2003

The moral rights of the authors have been asserted
Database right Oxford University Press (maker)

First published 2003

All rights reserved. No part of this publication may be reproduced,
or transmitted in any form or by any means, electronic or mechanical,
including photocopying, recording or by any information storage and
retrieval system, without permission in writing from Oxford University Press.
Enquiries concerning reproduction outside the scope of the above should be
sent to the Rights Department, Oxford University Press, at the address above

You must not circulate this book in any other binding or cover
and you must impose this same condition on any acquirer

ISBN 019 566119 2

Typeset in Goudy
by Eleven Arts, Keshav Puram, Delhi 110 035
Printed in India by Sai Printopack Pvt. Ltd., Delhi 110 020
Published by Manzar Khan, Oxford University Press
YMCA Library Building, Jai Singh Road, New Delhi 110 001

BL
1124.56
.G74
2003

070703-2196T8

In memory of
my father Yoel
and
my mother Orah

Preface

Upanishadic tales are about men and women in crisis, awakened to their inferiority, the painful consciousness of a gap between (their own) lesser selves and an elusive better self. These men and women are described by the early Upaniṣads as suffering metaphysical (or ontological) weakness (analytically distinct from other sorts of inferiority). Emerging from crisis, they close the gap between lesser and better selves, moving towards excellence and confidence. Thus, occasions for experiencing 'ontological inferiority' and its transcendence are the crux of Upanishadic storytelling.

The presence of a higher self is healthy as well as tormenting for Upanishadic characters; without such a presence there would be no 'Upanishadic pain' or consequent motion to alleviation through insight. Alert and attentive to their crisis, Upanishadic recipients of knowledge seek 'immortality' through understanding the better self (*ātman*). Indeed, the Upaniṣads suggest that identification of the great, absolute *ātman* as one's self is the most significant moment in life. This moment or event of self-identification is one of great joy, marking an end of suffering; under certain circumstances, the metaphysics of the *ātman* heals. Upanishadic storytelling is, thus, about the transition from inferiority to excellence by virtue of therapeutic knowledge.

The attention of the world has focused on the great and abstract philosophy expounded in the Upaniṣads, while the stories themselves have been under-read. Scant attention has been paid to the context of transmission of knowledge, culminating in dialogues between teachers and others. The stories' references to characters in need of knowledge are usually viewed as mere background to the exposition of liberating, abstract knowledge. Yet the recipients of knowledge—ordinary yet sensitive men and women in crisis—are, in my view, also of hermeneutic interest and import. Indeed, the essential Upanishadic narrative is about them. The complacement householder awakened to an ascetic's superiority (and his 'own' inferiority); the childless wife (bereft thereby of immortality); the young student afflicted by desires of many kinds (also for his teacher's wife); the boy insecure because of his unknown father's identity; the vain son perplexed by his father's riddles; the scholar full of scholarly yet impotent 'knowledge'; the boy insistent on his perception of mortality—these are some of the characters who live in the stories. All benefit from knowledge of the better self, and the Upanishadic teaching of the *ātman* is transmitted to them, and for their sake.

The healing potency of 'knowledge of the better self' is the theme underlying this essay. One should not, however, ignore the circumstances. Under certain circumstances, metaphysics heals; under others, it may not. Common readers like ourselves (whom I have termed twenty-first-century 'margin-dwellers') may not benefit as much as a Śvetaketu of Jānaśruti from teachings of the self expounded by sages such as Uddālaka, Raikva, or Yājñavalkya. One may not be sufficiently educated or open to absorb the Upanishadic message of the self, or one may lack the right teachers. It is a fact, then, that reading the Upaniṣads does not heal. A sense of helpful disease is, however, available, as readers of the Upaniṣads experience the vague yet real attraction of an underground, better self.

The Upanishadic tales and philosophy have not evolved in a vacuum; these texts reach us from the Indian subcontinent of 2,700 years ago. Although a link between historians' accounts and the momentous awakening of individuals is far from obvious or clear, scholars of Indian history do tell of major changes which occurred and accumulated in the Indian subcontinent at that time. It was a period of changes in technology (extensive use of iron), commerce and economy (surplus of food and other commodities), mobility, transportation, and urbanization. It was also a time of prosperity, growing leisure, and the rise of individualism in the wilderness, beyond the boundaries of family and village. Such changes affected family, village, culture, and values. Thus, available scholarly descriptions provide some background for reading the Upanishadic tales as narratives of crisis. The transition from an Old Vedic, offspring-based notion of immortality to the new Upanishadic teaching of the *ātman* is reflected in the biographies of the characters of the Upanishadic stories.

It is, thus, in correspondence with historians' descriptions of old India that an overwhelming sense of inferiority emerges in the lives depicted in the early Upaniṣads. Expressing this emergence of inferiority are tales of a seemingly successful householder (Jānaśruti), a childless woman (Maitreyī), a rejected student (Upakosala), a boy awakened to the mortality permeating health and the body (Naciketas), a great scholar (Nārada) painfully conscious of the futility of 'mere accumulation of information', and others at similarly critical junctures. Vaguely aware of a different horizon or mode of being or self, these characters—whose names and modes of inferiority are recorded in the early Upaniṣads—seek to know the better self by means of others, the teachers, gurus, Upanishadic therapists, as it were. Since knowledge of a better self in the Upaniṣads means *becoming* the better self, the transmission of knowledge implied in these documents must be assumed to be an arduous process (hence my characterization of their 'therapeutic' aspects).

The essential structure of the early Upanishadic text is of two parts: a tale of individual crisis and the exposition of knowledge (philosophy), capable of its alleviation. Consider, for example, the familial crisis instigated by a husband's decision to end his life as a householder and become a *sannyāsin*. Upon Yājñavalkya's decision to leave home (and village and his two wives), his learned wife, Maitreyī, asks him whether property will make her immortal (*amṛta*). It will not make you immortal, says her husband. She then asks him to tell her what he knows, and he does (telling her of the one *ātman* which is everything). This is simple enough, yet the story invites further questions. What is the connection between Maitreyī's preoccupation with 'immortality' (*amṛtatva*) and her husband's departure? In my disposition to look for crisis, I add something here, namely, Maitreyī's (and Yajñavalkya's) childlessness. The story itself as well as the well-known importance of offspring in defining Vedic immortality, strongly points to Yājñavalkya's childlessness. Had she had sons, Maitreyī would not be as vulnerable to her inner promptings on immortality. Her husband leaving her without sons in eighth-century BC India is her opportunity to become personally and metaphysically inferior with regard to immortality, and this makes her a proper Upanishadic heroine.

As this book is about re-reading the Upanishadic tales of crisis, it naturally focuses on men and women of the 'lesser self.' (Indeed, readers may often share lesser-self-existence with the Upanishadic heroes.) I see now that I had not well estimated the magnitude of the project I had in mind. Explication of the connection of crisis *and* knowledge requires not only thinking in depth on both (the condition of the afflicted Upanishadic hero or heroine and the knowledge of *ātman*), but also describing the transition from one kind of existence to another (higher, as it were). Tracing the connections of personal Vedic crisis with the emergence of the better-self-existence (or metaphysics) seems to me a worthwhile enterprise, but the present essay does not actually meet this challenge.

As this book evolved, its true theme became the importance of the storytelling as 'tales of crisis', significant in themselves, revealing of Upanishadic predicament, worthy of attention. Yet I have been told that this book, while endowed with imagination and insight, is somewhat frustrating, not fulfilling its promise. The idea, they say, is good: to read the Upanishadic stories as closely as possible and to re-read the familiar yet 'ineffective' messages (of the pure self—*ātman*—or the absolute Brahman) as significantly and strictly embedded within the storytelling. The Upanishadic messages do not reach our souls forcefully enough; indeed, we are hardly an engaging audience for Uddālaka's *tat tvam asi*. He is addressing his eighth-century BC son-in-crisis, Śvetaketu, not us. However, closer reading of the Upanishadic narrative is a move toward shrinking the gap between the Upanishadic audience-in-crisis and us (twenty-first century readers). But the contextual metaphysics I am seeking is elusive, and the connection of story and overt philosophical message remains more obscure than I had hoped. The resonance of the deeper-than-life self is not audible enough within the story. The tales of Jānaśruti's eruption of inferiority-consciousness, Maitreyī's childlessness (non-eternity), Śvetaketu's inflated ego and consequent crisis, Nārada's sorrow, Satyakāma's doubt, etc., do not tangibly—organically—*contain* the Upanishadic teaching of the *ātman*. Yet, I believe that the pursuit does make sense, and is at least forcefully expressed. I hope I have created an opening for further thought and investigation of the connection of crisis to knowledge—and storytelling to healing.

Acknowledgements

This book was conceived with the help and encouragement of many, only three of whom I will mention here.

Prof. David Shulman has created much of the horizon of my efforts. His life-long interest in the powers of language and his insistence on the metaphysical dimension of any linguistic expression find an echo here. In his relentless wonder at the big and basic questions in life and scholarship, he has made many, myself included, more serious and creative. I would have done nothing in Indian studies without his inspiration and attentive interest. One afternoon, by a casual question (why indeed have Indologists under-read the Upanishadic stories?), he sealed my intention to write about these tales.

Dr Nita Schechet, in her independent and penetrating reading of poetry and narrative, has been midwife to this book. She articulates a more personal expression in scholarly matters, and thus has made me—a somewhat dry writer—freer in my interpretations.

My son Yoel inserted the diacritical signs and compiled the bibliography with commitment and skill, and out of love he has driven me to finish this book. I am grateful to him and to others who helped me overcome my love for 'idling.'

Abbreviations

Bhagavadgītā	BG
Bṛhadāraṇyaka Upaniṣad	BU
Brahmasūtra	BS
Brahmasūtrabhāṣya	BSBh
Chāndogya Upaniṣad	ChU
Chāndogya Upaniṣad Bhāṣya	ChUBh
Kaṭha Upaniṣad	KU
Kauṣītaki Upaniṣad	KauU
Manusmṛti	MS
Muṇḍaka Upaniṣad	MuU
Ṛg-Veda	RV
Śatapathabrāhmaṇa	SB
Taittirīya Brāhmaṇa	TB
Taittirīya Upaniṣad	TU

Contents

Preface		vii
Acknowledgements		xi
List of Abbreviations		xii
Chapter 1:	On Good-Enough Reading of the Upaniṣads	1
Chapter 2:	Personal Crisis and Contextual Metaphysics: Reading the Under-Read Stories of Upakosala K. and Satyakāma J.	25
Chapter 3:	Marginality and Great Moments: Contextual Metaphysics in the Story of Maitreyī	57
Chapter 4:	Under-Reading Multiple Vocality: The Case of the Good Boy and the Angry Father	80
Chapter 5:	Colourless Words or Contextual Hermeneutics: the Visible and Invisible Narratives of *Chāndogya* 6	101
Epilogue:	Storytelling and Fearful Self-Understanding	130
Bibliography		138
Index		140

ONE

On Good-Enough Reading of the Upaniṣads

The origin of this essay lies in my wonder at the redeeming potency of Upanishadic speech, in particular its 'metaphysics'. Indeed, the Upanishadic composer's self-understanding never tires of expressing itself in this vein; the protagonists of 'knowledge' promise their audience power, bliss, immortality, the shore beyond sorrow. Thus, it is widely believed that the Upaniṣads provide a particularly forceful type of knowledge, a 'therapeutic metaphysics', as it were. How can we wisely focus on such Upanishadic promises, explicit and implicit?

In my view, the connection between storytelling and abstract instruction of the self is a most intriguing, most promising area of research into the meaning and power of Upanishadic speech. Reading connections between stories and their import is obviously anchored in the very nature of the structure of the Upanishadic text. The Upanishadic story is never offered for its own sake. Nor are the sublime messages delivered context-free. Instead, they are woven into a story told through individual speakers who seek something—alleviation or resolution of crisis. The Upanishadic storytelling highlights the difficulty and paradox in maintaining Vedic otherness in its entirety (as a holistic, uniform attitude).

To present such an approach in this essay entitled 'Good-Enough Reading', let us consider an Upanishadic section which contains a story and a message and reflect on differences in approach. The section selected here is the Jānaśruti story from the *Chāndogya Upaniṣad* 4.1, a section containing a complex story and the metaphysical knowledge 'embedded within'.

Jānaśruti is a very generous householder, building and sustaining villages and *aśrams*. Apparently, he feels good about himself, saying: 'Everywhere people will be eating my food.' But one night, two eloquent geese (*haṃsa*) fly over his head. He listens to their talk, and lo! They are talking about him! Of all things in the world, these two geese are conversing about Jānaśruti. One of the geese seems to echo Jānaśruti's good feeling about himself. He admonishes the other goose to fly cautiously over Jānaśruti's head, since the celestial light (*jyotiḥ*) emanating from it may burn them. However, the other goose does not recognize Jānaśruti's greatness and does not even seem to not know who Jānaśruti is; thus, he enquires about Jānaśruti, spoken of as if he were Raikva-of-the-cart. But the first goose who is aware of Jānaśruti's greatness and splendour does not know of Raikva, and is curious about who Raikva is. The second goose replies with a metaphor from the game of dice—as the lesser throws are inferior to and go to the winning throw called *kṛta*, so everything (*sarva*) goes to him (Raikva). All virtuous deeds of creatures are attributed to him (*yathākṛtāya vijitāyādhareyaḥ saṁyanti evam enaṃ sarvaṃ tad abhisamaiti yat kiñca prajāḥ sadhu kurvanti*). And the road to completion, confidence, and excellence is made explicit: whoever knows what Raikva knows becomes like him. Śaṅkara adds: he becomes like the winning throw (*kṛta*) in the game of dice.

Overhearing the geese's talk is apparently the most momentous event in Jānaśruti's life. He is invaded, even tormented by a new sense of inferiority. From a somewhat complacent, self-fulfilled, respectable householder, he is transformed into an anxious person subject to self-conflict urgently seeking remedy for his inferiority. A successful life 'in the world' seems to lose its significance as a sense of inner inferiority suddenly comes to the fore. Indeed, the correspondence between the outer and the inner aspects of Jānaśruti's life is remarkable. Up to the crucial moment of overhearing the geese's conversation, Jānaśruti is satisfied and does not seek further 'knowledge'. The geese's conversation, prompting Jānaśruti's inner turmoil and crisis, may well reflect Jānaśruti's potential and long latent feelings of inferiority. Why else did Jānaśruti hear the geese talking about himself and not about somebody else? The narrator could well have told Jānaśruti only about Raikva, without so blatantly injuring Jānaśruti's self-confidence. But the apparently external, night-time event is like a dream, exposing fundamental inner truths. Jānaśruti's sense of inferiority—well concealed by his successful projects 'in the world'—is exposed, materialized in and through the avian dialogue. While Jānaśruti's property, energy, and people's admiration for him suggest a stable, well-established sense of confidence and power, the talk in the dark exposes the truth of a divided, incomplete, precarious inner self.

The storytelling imparts the allure of a contest most pertinent to the clarification of Jānaśruti's inner division. We have already noticed that the two geese know different things; one of them knows Jānaśruti, the other knows Raikva. The one who knows Jānaśruti does not know Raikva, and the one who knows Raikva does not know Jānaśruti. Thus, not only is this 'division of knowledge' made explicit in the story, it is a necessary condition for inner dialogue (each goose asks the other about his hero—a new object of knowledge). It is a tense, essentially competitive dialogue. Even before the metaphor of the game of dice is ushered into the narrative, the geese compete over who knows more. Their conversation is thus expressive not only of bifurcation and conflict, but also of dialogue. Viewing the two geese as Jānaśruti's 'inner voices', we may conclude that Jānaśruti's inner being is painfully split. But this split is structured. One part is superior yet present potentially only; the other is actual and inferior. Obviously, the one who knows Raikva knows better and is thus superior to the one who knows Jānaśruti. Jānaśruti, indeed, draws the same conclusion. Heeding Raikva's superiority as the message of the dream-like night-time conversation, he is driven henceforth to bridge the gap between his poor and better selves. At the end of the road he is about to receive 'knowledge' which subsumes his suffering, a point we shall revisit later in this discussion.

The Upaniṣads, invariably considered foundational to Indian civilization, have been approached in many ways. Eager seekers, memorizing pandits, imaginative poets, objectifying historians and other scholars, Indian nationalists, other-worldly spiritualists, well-intentioned Western reformers, and many others have all created their own contexts for reading the Upaniṣads. My own introductory question—whence, in the Upanishadic self-understanding, the power of metaphysics/knowledge—constitutes another such context of interaction with the ancient texts. Assuming that the Upaniṣads testify to the transforming power—hence its fateful relevance—of the 'knowledge' transmitted by its protagonists, it already establishes its own criterion of 'Good-Enough Reading'. Such reading (always also an interpretation) should throw light on the nature of the force of Upanishadic knowledge (rather than merely clarify its 'contents'). Rooted in this context of 'Good-Enough Reading,' this study of the force of consciousness is primarily focused on the relevance of storytelling to the nature and power of Upanishadic knowledge (*jñāna*). The present approach is essentially oriented toward truths repeatedly, explicitly, relentlessly expressed in the Upanishadic texts. It thus raises questions undeniably springing from the 'texts themselves'. However, though undertaken in recognition of intentions embedded in the texts, the present exploration is of course not context-free. None of the voices mentioned above (the pandit's, historian's, seeker's, Orientalist's, etc.) speaking for Upanishadic meanings is—by definition—value-free or context-free. All are most valuable, enlightening voices emanating from particular, special, and genuine perspectives, enriching subsequent readings. The variety of attitude and pitch of these voices is remarkable, even within the more scholarly community. This is indeed the beauty of great texts—the attraction they hold for so many different minds and the variegated responses they elicit. While listening attentively to context-laden pitches and sounds invites a better reading of such great texts, I cannot represent the immense variety of the interesting voices commenting on the Upaniṣads. And yet, it is essential to characterize some of the major approaches in the reading of these sacred Indian documents as part of the study of the issues raised above—the nature and force of Upanishadic knowledge.

Having attentively reviewed documented contacts with the Upaniṣads in various reading-experiences of these texts, I can roughly classify four generic modes of contact with the Upanishadic texts into categories I would call traditional, existential, secondary, and good-enough readings. The criteria for classification are the articulation of the significance of Vedic otherness and the attitude towards the power allegedly generated in course of the textual interaction.

Traditional Reading is committed to the extreme, essential otherness in Vedic action and presence, as well as to its revolutionary power. Existential Reading denies Vedic otherness but is open to its power and relevance in human life. Secondary Reading refers to the nature of Vedic otherness and power indirectly, mostly as dimensions relevant for others (not for the Secondary Reader himself), thus tacitly denying Vedic otherness and force. Good-Enough Reading incorporates elements from the three other approaches, while seeking in particular to make sense of the Upanishadic experience of the force embedded in Vedic speech, especially as this experience is reflected in Upanishadic storytelling.

Traditional Reading is based on cautious preservation of Vedic otherness, consequently resisting tendencies to ground interpretations of the Upaniṣads in normative ('human') experience. Quotation, philological (and etymological) analysis, and meticulous (and strictly ruled) textual study[1] are exemplary expressions of this approach. Along with its commitment to the otherness of Vedic sources, Traditional Reading accepts

[1]Identification of the main subject of a passage (*prakṛta*, *prakāraṇa*) and the meaning and reference of 'complementary sentences' (*vākya-śeṣa*) are among the more important rules of textual analysis followed by Śaṅkara.

the fateful importance of the correct understanding and absorption of Vedic intentions. Such intentions lead to new openness ('knowledge'), to constant recognition of the field of the Absolute (Brahman) ground of this world. Since Vedic intentions are embodied in speech, the clarification of the objective meaning of Vedic words is the means to reach openness to the ground—understanding of Brahman.[2] Combining Vedic otherness with the abysmal difference between the phenomenal world (*saṁsāra*), the Absolute and liberation (*mokṣa*), the Traditional Reader correspondingly dissociates himself from the text in order to listen to it (namely, to the other's intentions). While any reading involves self-forgetfulness of a kind, the Traditional Reader aspires to self-forgetfulness in a more literal, seemingly more complete mode of reading. And yet, Vedic (allegedly extreme)—non-human—otherness is essentially elusive. Thus the Traditional Reader's very contact with the scorching Vedic otherness is loaded with hermeneutic paradox and ambivalence.

For the Existential Reader, moments of self-forgetfulness, while necessarily being an integral part of reading, are less conspicuous and encouraged. Existential Reading is personal, grounded in experiences and meanings available to the reader in course of the interaction with the Upanishadic text. Substantially ignoring Vedic otherness, such generic reading is essentially a projection of normal experience on to the Vedic text. Man's inner being ('heart') is the source of interpretation and the measure of the validity and value of the Veda. Fortunately, the Existential Reader would assert that the intentions of the Upaniṣads are fully commensurate with the spiritual experiences in our own age. Supreme joy (*ānanda*), for example, is a symptom of contact with the ultimate reality, according to both the Upaniṣads and ordinary life.

> In the Upaniṣads we find the note of certainty about the spiritual meaning of existence ... they aver that through our joy we know the reality that is infinite, for the test by which reality is apprehended is joy. Therefore in the Upaniṣads *satyam* and *ānandam* are one. Does not this idea harmonize with our everyday experience?[3]

This approach in the interaction with the Veda (Upaniṣads) presupposes the connectedness of (our) phenomenal world with the happy though

[2]See for example, BSBh 1.1.2: *vākyārtha-vicāraṅādhyavasāna-nirvṛtta hi brahmāvagatiḥ* ['Knowledge of Brahman is obtained by ascertainment of the meanings of (Vedic) speech through investigation'].

[3]R. Tagore, Appendix A, in S. Radhakrishnan (ed.), *The Principal Upaniṣads*, George Allen & Unwin Ltd., 1953, pp. 941–2.

'empty' universe of *mokṣa*. The condition of *mokṣa* is—for the Existentialist—inherently connected to man's condition in *saṃsāra*, *mokṣa* being a state of resolution of problems essential to the samsaric condition.

The third approach is the Secondary/Scholarly Reading of the Upaniṣads. Neutralizing existential drives and personal experience (in the spirit of value-free scholarship), seeking to clarify objective meanings (to dissociate oneself from the text and to understand it), it is significantly close in spirit to Traditional Reading. Indeed, Secondary Readers in principle and invariably pay homage in speech and deed (interpretation) to the great authorities of Traditional Reading (Śaṅkara, Vācaspati, etc.). However, unlike Traditional Reading, the Secondary approach is not committed to the primary and direct relevance of the Upanishadic teachings (documents). In general, most of the exponents of this approach explicitly ignore possible spiritual value inherent in the Upanishadic documents. While the Upaniṣads are accorded immense historical value as being central to Hindu self-definition, reference to their spiritual value is often beyond the overt sensibilities and aspirations of the Secondary Reader. Such a reader notices the Upanishadic insistence on 'forceful metaphysics', but is not curious about its underlying dynamics and meanings. Some (such as S. Radhakrishnan) are close to asserting personal commitment to the spiritual truths in the Upaniṣads. Others express their somewhat distant, even negative appraisal of these texts, triggering hostile responses from Existential Readers who cannot stand the objectifying, seemingly condescending spirit of Secondary Reading. Rabindranath Tagore's attack on Gough's *The Philosophy of the Upanishads and Ancient Indian Metaphysics* is an example. He complains that the 'lack of sympathy and respect displayed in it for some of the most sacred words that have ever issued from the human mind, is amazing'.[4]

The fourth approach to an interaction with the Upaniṣads is what I would call Good-Enough Reading. It is essentially an ecclectic approach,

[4]Ibid., p. 940. A.E. Gough's *The Philosophy of the Upanishads and Ancient Indian Metaphysics* (1882) is indeed a most negative representation of the 'Indian mind'; it ends in the following paragraph: 'Such as they are, and have been shown to be, the Upanishads are the loftiest utterances of Indian intelligence. They are the work of a rude age, a deteriorated race, and a barbarous and unprogressive community. Whatever value the reader may assign to the ideas they present, they are the highest produce of the ancient Indian mind, and almost the only elements of interest in Indian literature, which is at every stage replete with them to saturation' (p. 268). Thus Gough sums up the philosophy of the Upaniṣads: 'It is no aspiration and energy towards the true and the good, but only a yearning for repose from the miseries of life. Yet it is the highest product of the Indian mind' (p. 267).

incorporating elements of all the other approaches, yet adhering to none. In its fundamental interest in the force of Upanishadic knowledge and its transformative potency of the transition from *saṃsāra* to *mokṣa*, Good-Enough Reading focuses on man's samsaric as well as post-samsaric condition. It suggests that Upanishadic storytelling may be viewed as accounts of transformation, reflecting pre-liberation existence and the nature of liberating knowledge. Rejecting the constricting rigidity of the Traditional Reading's conflictual contact with the transcendental otherness, Good-Enough Reading accepts the Traditional commitment to the idea of 'forceful knowledge', to the force and value of Upanishadic teaching. It is thus close to the Existential spirit, open as it is to the Upanishadic challenge to prevailing conditions of 'ignorance.' Good-Enough Reading is also open to information produced by the circles of Secondary Reading, seeking above all to explore the Upanishadic narrative as descriptions of 'contexts of transformation' (such as personal crisis and its resolution by 'knowledge').

The Upaniṣads are among the more difficult texts for readers to share. These are Indian documents, originally composed in Sanskrit, 'mystical', charged with distance and otherness, conceived by the sages of old immersed in harsh disciplines and meditative practices, well-versed in the methodology of ancient Vedic culture. It is possible that they used drugs and experienced corresponding states of consciousness we are unaware of; they often addressed intimate, transformative situations. Reading these texts, having a share in them, seems an almost impossible task.[5] And yet, knowledgeable persons do feel in these texts universal echoes of sublime philosophy, such as that of Plato, Spinoza, or Kant, which refer to the Brahman—the dangerously charged subjective power beyond decay and death, the unperceivable essence present everywhere. While most readers may not agree with Schopenhauer's proclamation that the Upaniṣads are the solace of life and death, they nevertheless find their own modes of meaningfully engaging with the Upaniṣads. The Upanishadic documents have been and continue to be vital to Indian culture, and the West too has long responded to the Upaniṣads directly (as a source of inspiration) as well as indirectly (in recognition of their importance for others).

The Upanishadic text is invariably complex and versatile. Confining

[5]Even within the Upanishadic texts themselves there is recognition of a difficulty in understanding the transforming Upanishadic message. The Upaniṣads repeatedly tell us that understanding them is very difficult. 'Tell me the Brahman directly and unambiguously perceived', says one of the sages eager to know the self (BU 3.4–5). Thus, a measure of self-reflexivity is present in the Upaniṣads.

ourselves to the earlier Upaniṣads—the focus of this essay—we notice the overwhelming variety of voices within them. In the spirit of the older Brahmaṇas and Āraṇyakas, the Upaniṣads contain sections dealing with sacrifice and ritual,[6] the use of holy and potent syllables,[7] meditations transforming man into the object of his desire and the focus of his meditation. There are also stories of sages, ascetics, disciples, householders, fateful intellectual contests and debates. Sublime philosophical messages are invariably embedded within narratives of personal and interpersonal conflicts and crises. 'Upanishadic knowledge' is offered as a remedy for man's sorrow in *saṃsāra*.

The mystical and philosophical assertions of the Upaniṣads have been overwhelmingly significant for India and the West. Some, like Hegel, have seen inferiority manifested in the fundamental Upanishadic references to the abstract identity of Self and Absolute. Others have admired the articulations of the essential identity of the innermost core of the human being with the Real. Surprisingly few have addressed Upanishadic storytelling in its connection with the abstract teachings embedded in these stories. The relationship between story and meaning has most often been regarded as mechanical, the story providing external circumstances, a mechanism for conveying the liberating knowledge. Śaṅkara's theory of the function of the Upanishadic story (BSBh 3.4.23–24) expresses the basic perception in Traditional Reading of the relationship between story and metaphysics. In his view, the Upanishadic narrative stands in proximity to the highest, impersonal metaphysics solely in order to motivate the audience to listen. Though the personalities involved in the circumstances of knowledge transmission seem in dire need of instruction (a need inviting empathy, yet contrary to the maintenance of otherness), they are invariably doomed to secondary, even marginal significance.

What I deem Traditional Reading is based on and exemplified by Śaṅkara's interpretation of the Upaniṣads. Śaṅkara's experience of Vedic presences, his coherent and well-integrated interpretations of the Upaniṣads and other pillars of Vedāntic culture (such as the *Brahmasūtra* and the Bhagavadgītā) have become constitutive of the Vedānta tradition.

Śaṅkara experiences the Veda as the other (non-human), autonomous

[6]See, for example, the opening section of the *Bṛhadāraṇyaka Upaniṣad*, dealing with the symbolism (correspondence) of the various members of the horse. This section is an exact replica of the section from *Śatapathabrāhmaṇa*.

[7]See, for example, the discussion of the *Udghīta* syllable in *Chāndogya Upaniṣad* and the numerous references to the nature and power of the syllable Om.

(*svatantra*), scientific, inexplicably benevolent, and multi-vocal. The Veda is a valid means of knowledge (*pramaṇa*) and it teaches one essential, life-giving truth—of the underlying factual unity of all. The ultimate voice of the Veda is clear in the 'great sayings' (*mahāvākya*) such as 'Thou are That' (*Tat tvam asi*), 'I am Brahman' (*ahaṃ brahmāsmi*), and 'This self is the Absolute' (*ayam ātmā brahma*). Such utterances are the purest sounds of Vedic perspective and otherness. Corresponding to the nature of the Veda as foreign and other is the experience—often shifting yet primary—of the phenomenal world (known by human faculties) as distinct from the Absolute. These aspects of Traditional Reading provide a well-integrated approach and mode of engaging with the Upaniṣads.

Traditional Reading seeks understanding of the literal meaning of the text, pointedly resisting empathic impulses. Such reading avoids research into the Upanishadic heroes' lives and experiences. Indeed, Traditional Reading adds conspicuously little to our knowledge of the conditions in which Upanishadic knowledge is promulgated. Living in the presence of Vedic otherness, the Traditional Reader does not make use of much that he knows and experiences. Take, for example, Uddālaka's reluctance to teach the five householders who approach him for instruction of the *ātman*. According to *Chāndogya Upaniṣad* 5.11.3, he refuses to teach them, thinking 'I will not answer everything (they may ask)', (a seemingly unique reference to a man thinking to himself.) A Traditional Reader like Śaṅkara, most probably experienced in the practice of transmitting 'Upanishadic knowledge' of the self, could add much to our understanding of Uddālaka's 'state of mind' (or the meaning of his refusal to teach). But alas! The Traditional Reader says nothing of Uddālaka's apparent fears and predicament, nor of the meaning of his solution (relegation of instruction to another teacher). Indeed, empathic reading is hardly discernible throughout the Traditional Reader's reading of the Upaniṣads. No reference is made to Maitreyī's condition of apparent childlessness, to Yājñavalkya's relationship with Uddālaka, to Raikva's puzzling decision to instruct Jānaśruti by virtue of Jānaśruti's daughter's face, to Pratardana's recalcitrant behaviour regarding Indra's promise of a boon, etc. This silence about all-too-human components of the Upanishadic narrative is commensurate with the abysmal Vedic otherness explicated as non-human (*apauruṣeya*).

However other and non-human in terms of origin, Vedic speech is communicative, accessible to man. Its otherness is thus mitigated; though other, it may be understood for the benefit of man, by virtue of inexplicable, unfathomable benevolence. Thus Śaṅkara says on BS 1.2.2: 'Although in the Veda which is not the work of man no wish in the strict sense can be expressed, there being no speaker, still such phrases as "desired to be expressed", may be figuratively used on account of the result, that is, (mental) comprehension.'[8] Śaṅkara experiences the Veda as a hierarchical, complex, pedagogical action aimed at redeeming man's understanding.[9] Vedic speech is unified, coherent, essentially a single speech-act culminating in *mahāvākyas* such as *Tat tvam asi*. In this approach, *Tat tvam asi* is a statement expressive of the entire Veda rather than an address of a father (Uddālaka) to his somewhat resistant, even hostile son (Śvetaketu). Though singularly focused on the metaphysical, scientific voice[10] of the

[8]Śaṅkara's BSBh 1.2.2. (Thibaut's translation, in the *Vedānta-Sutras*, Motilal Banarsidass, 1904, p. 110).

[9]However, recognizing the coherent, unitary nature of the Veda—oriented exclusively, relentlessly toward the transmission of knowledge—also implies Vedic multivocality, for the Vedānta-documents (the Upaniṣads) apparently speak of many things. How to conceptualize the enormous scope and variety of Upanishadic speech? Three voices of the Veda are patently referred to by Śaṅkara; the voice of command, injunction (*vidhi, codanā*), the voice of praise (*arthavāda*), and the voice of knowledge. The foreground, main voice is the voice of Vedic knowledge. The other two voices are supplementary, supporting the main voice in various ways. To which of these does the storytelling belong? It is *arthavāda*; a complementary, supportive, motivating statement of lesser force and not as fully true as the great scientific sayings of the Veda (*mahāvākyas*). Most important in Śaṅkara's mode of reading the Upaniṣads is his insistence on the independence (*svātantrya*) of 'knowledge.' Śaṅkara's main assertion is that liberating knowledge (*jñāna*) is independent of ritualistic action (karma). It is an assertion grounded in experience; the condition of liberation is inherently different from samsaric existence. It is incomparable in essence, not in degree. In terms of reasoning and verbal expression, Śaṅkara asserts that if liberation depended upon, or was produced by something else, it would not be eternal (as it is) (BSBh 1.1.4). *tatraivaṃ sati yathokta-karma-phaleṣv eva tāratamyāvasthiteṣv anityeṣu kaścid atiśayo mokṣa iti prasajyeta. Nityaś ca mokṣaḥ sarvairmokṣavādibhirabhyupagamyate, ato na kartavyaśeṣatvena brahmopadeśo yuktaḥ. Api ca 'brahma veda brahmaiva bhavati'* (MuU 3.2.9). Śaṅkara describes the abysmal difference between *saṃsāra* and *mokṣa* also in terms of the bodilessness of *mokṣa*; 'What is called *mokṣa* is but this (state of) bodilessness' (*tad etad aśarīratvaṃ mokṣākhyam*). 'All of which passages establishes the fact that that so-called release differs from all the fruits of action, and is an eternally and essentially disembodied state. Among eternal things, some indeed may be "eternal, although changing" (*pariṇāminitya*), the idea of whose identity is not destroyed, although they may undergo changes; such, for instance, are earth and the other elements in the opinion of those who maintain the eternity of the world, or the three *guṇas* in the opinion of the Sāṅkhyas. But this (*mokṣa*) is eternal in the true sense, that is, eternal without undergoing any changes (*kūtasthānitya*), omnipresent.'

[10]Śaṅkara often emphasizes the nature of the highest pitch of the Upaniṣads as a factual, scientific expression, solely dependent on the object of reference, thus entirely 'objective'. See, for example, BSBh 1.14.

Veda, the Traditional Reader—Śaṅkara—does not ignore the Vedantic story altogether. He assigns the story functions and value, and is sometimes intrigued by and creative in his contact with the story. For example, in his commentary on the Upanishadic story of Uddālaka and Śvetaketu, he asks questions concerning the relationship of father and son; he raises a new, startling question: why does the father not initiate and teach his son? (Śaṅkara apparently perceives this as a natural possibility. His answer, however, avoids exploring the implications of his question. 'The father was away,' he says.)

As a generic mode of engaging with the Upaniṣads, Existential Reading is described here as contrary to the Traditional, Śaṅkarite, Reading. The Existential Reader attends to his own needs, identifications, hopes and experiences, and in accordance with these he interprets the Upaniṣads. A. Schopenhauer, speaking of the Upaniṣads as the solace of his life and death, avoids much Upanishadic otherness; although unable to read the Upaniṣads in Sanskrit, he feels close to these ancient—apparently foreign—texts, identifying in them his very own, most intimate thoughts.[11] Similarly, H. Zimmer emphasizes the need of the West to make Indian wisdom its own. Unlike Hegel, he suggests that Indian philosophy manifests a more advanced stage of man's spiritual condition than that of the West. 'We of the Occident are about to arrive at a crossroads that was reached by the thinkers of India some seven hundred years before Christ.'[12] However, Zimmer says that while the Indians formulated problems crucially pertinent to our West, the latter 'cannot take over the Indian solutions'. Thus, openness to the other's existence and import is—somewhat paradoxically—defined as contingent upon the recipient's universe of experience and meanings. 'We shall therefore have to follow the difficult way of our own experiences, produce our own reactions, and assimilate our sufferings and realizations. Only then will the truth that we bring to manifestation be as much our own flesh and blood as is the child's its mother's' (p. 2). In this vein, the Existential Reader defines the Upanishadic sages in his own (Western) image, as 'freethinking', 'creative philosophers', 'pioneer intellectuals', etc.[13] Perceived as a freethinker or creative philosopher, the ancient Upanishadic sage is otherness-free, as are his ideas.

On the Indian side, Debendranath Tagore (1817–1905) is an exemplary exponent of the existential ideology and approach to reading the Upaniṣads.

[11] H. Zimmer, *Philosophies of India*, Meridian Books, 1957, Ch. I.
[12] Ibid., p. 1.
[13] Ibid., p. 355–6.

In accordance with the main orientation of Neo-Hinduism, Tagore's emphasis is on the role of experience as the foundation of religion, in particular with respect to reading texts such as the Upaniṣads:

Brahma reigned in the pure heart alone. The pure, unsophisticated heart was the seat of Brahmaism. We could accept those texts only of the Upaniṣads which accorded with that heart. Those sayings which disagreed with the heart we could not accept.[14]

Tagore's version of Existential Reading shares with Traditional Reading the assumption that an original (Vedic) intention does exist in the Upaniṣads. However, he considers receptivity to the other's intentions (text) in terms of pure heart rather than scholarship, textual analysis, erudition, and 'knowledge'. Thus, he seems to locate himself at the vantage point of the original creator of the Upaniṣads while recognizing the dire need for help from the outside to establish his own autonomy over the sacred documents:

Debendranath places himself in the position of a 'seer' and attempts to personally realize and reactualize what is documented in the Upaniṣads—at least as far as they are true and acceptable: 'Thinking thus, I laid my heart open to God, and said: "Illumine Thou the darkness of my soul." By His Mercy my heart was instantly enlightened. ... Thus by the grace of God, and through the language of the Upaniṣads, I evolved the foundation of the Brahma Dharma from my heart.'[15]

Debendranath's reading of the Upaniṣads is thus mediated by his application to God. Reading seems to consist largely of purification of the heart rather than intense textual engagement. Debendranath, as the ultimate Existential Reader, explicitly opposes Śaṅkara's approach to reading the Upaniṣads.

... Debendranath quickly broke with the Vedānta philosophy of Śaṅkara, whose non-dualism appeared completely unsuited to the establishment of a new religious and social life. He concluded that he should replace Śaṅkara's commentaries on the Upaniṣads with interpretations of his own.[16]

Debendranath's son, Rabindranath Tagore, follows his father in his departure from Traditional Reading with what I call here Existential Reading. While not a scholar, an Existential Reader is capable of sensing the

[14]See W. Halbfass, *India and Europe*, SUNY, 1988, p. 223.
[15]Ibid.
[16]Ibid., p. 233.

essence of the Upaniṣads, for the more significant part of reading the Upaniṣads does not imply scholarship:

It is not enough that one should know the meaning of the words and the grammar of the Sanskrit texts in order to realize the deeper significance of the utterances that have come to us across centuries of vast changes, both of the inner as well as the external conditions of life. Once the language in which these were written was living, and therefore the words contained in them had their full context in the life of the people of that period, who spoke them. Divested of that vital atmosphere, a large part of the language of these great texts offers to us merely its philological structure and not life's subtle gesture which can express through suggestion all that is ineffable.[17]

Thus Tagore asserts that the Upanishadic texts are essentially poetic, exercising their impact by 'suggestion'. Given the concise richness of their language, how then can the Upaniṣads be understood at all? Tagore suggests that somehow, as a matter of fact, this is done: 'Generations of men in our country, no mere students of philosophy, but seekers of life's fulfilment, may make living use of the texts, but can never exhaust them of their freshness of meaning.'[18] Moreover, Tagore concedes that there is value in the 'bondage of grammar' for the seeker of freedom:

If in our language the sentences were merely for expressing grammatical rules, then the using of such a language would be a slavery to fruitless pedantry. But, because language has for its ultimate object the expression of ideas, our mind gains its freedom through it, and the bondage of grammar itself is a help towards this freedom.[19]

Yet Tagore's arguments are not conclusive. How are the Upaniṣads read by those 'seekers of life's fulfilment'? How do they understand the Upanishadic words? Conscious of the insurmountable difficulties inherent in the contact with ancient, inaccessible intentions, Tagore seems to make an inexplicable leap of Existential Reading, expressing a seemingly implausible idea: the Upanishadic message is independent of the Upanishadic language:

Though many of the symbolical expressions used in the Upaniṣads can hardly be understood today, or are sure to be wrongly interpreted, yet the messages

[17]S. Radhakrishnan (ed.), *The Principal Upaniṣads*, George Allen & Unwin Ltd, 1953, Appendix A, p. 939.

[18]Ibid., p. 940.

[19]Ibid., p. 943.

contained in these, like some eternal source of light, still illumine and vitalize the religious mind of India.[20]

Thus, Vedic otherness—innately inaccessible—loses its essential distance from human existence altogether.

In seeming contrast, Secondary Reading of the Upaniṣads is explicitly impersonal, in keeping with the spirit of the value-free tradition of Western scholarship, closely resembling in this respect Traditional Reading. However, Secondary Reading ignores the Traditional commitment to the value of *jñāna*. The Upaniṣads are 'documents', an object of scholarly interest, not necessarily of spiritual value or a *pramāṇa* with respect to the Absolute. Even Indian scholars such as S. Radhakrishnan, S. Dasgupta, and M. Hiriyanna do not cross the boundary between Scholarly/Secondary and Traditional Readings. Such scholars, while appreciating the Upaniṣads, cautiously maintain a distance between their scholarly selves and the documents they explain. Secondary Reading focuses on the abstract contents of the 'Upanishadic teaching', transforming it into philosophy. Thus M. Hiriyanna sums up the teaching of the Upaniṣads against the horizon of scholarly vocabulary and distinctions:

There are great, almost insurmountable, difficulties in deciding what exactly is the teaching of the Upaniṣads in certain important respects. This accounts for the emergence in later times of diverse schools of Vedānta, all of which claim to propound the Upanishadic teaching. It is clear, however, that the prevailing view in them is monistic and absolutistic. That is to say, they teach that the ultimate reality is one and only one.[21]

Eclectic in approach, Good-Enough Reading seeks to absorb and apply most of the sensibilities embedded in the Traditional, Existential, and Secondary approaches. Good-Enough Reading attempts a close reading of the entire text, and thus, in its own way, it implements the spirit of Traditional Reading technically expressed by notions such as *vākyaśeṣa* (relevant expressions elsewhere) and *prakṛta* (principal subject of the section). It also seeks to interpret the Upanishadic text in terms of the alleged power of Upanishadic metaphysics, seeking an understanding of the Upanishadic force in closer attention to and more acute perception of the context of Upanishadic assertions of metaphysics. In this approach, which I hope to exemplify in the chapters of this book, the context of

[20]Ibid., p. 940.

[21]M. Hiriyanna, *The Essentials of Indian Philosophy*, George Allen & Unwin Ltd., 1985, p. 19.

the speech-act expressive of the message is an important clue to its understanding. In thus emphasizing the relevance of Upanishadic storytelling in the interpretation of the 'Upanishadic philosophy', Good-Enough Reading promotes a 'contextual metaphysics', a move I will illustrate here by returning to our story of the householder Jānaśruti.

Jānaśruti's newly acquired suffering is not necessarily the same as that of other Upanishadic protagonists. The story of Satyakāma's life centres on his inner doubts concerning the identity of his father, and consequently his own entitlement to 'knowledge.'[22] Maitreyī's suffering is also different; she seeks 'knowledge' prompted by a sudden awareness of childlessness and mortality.[23] Nārada's is primarily a scholar's predicament, his conventional knowledge patently insufficient and unfulfilling.[24] A prevailing sense of inferiority is apparently common to the recipients of Upanishadic 'knowledge'. However, the myths are different, the sense of inferiority is differently articulated, and the articulation of Upanishadic 'knowledge' is also different. The challenge and aim of Good-Enough Reading now become apparent; how can one read, or experience, the transmission of 'knowledge' as an *event* in the Upanishadic myth? The myth of Jānaśruti in the *Chāndogya Upaniṣad* 4.1–3 unravels Jānaśruti's richly elaborated interior; Good-Enough Reading of this story seeks explication of metaphysical meaning within the context of the story of Jānaśruti's life.

In the *Chāndogya Upaniṣad* the second goose's description of Raikva through the metaphor of the game of dice in which the lesser throws are incorporated in the winning one (*kṛta*) is a reflection upon Jānaśruti's essential inferiority. The nature of the game is somewhat obscure; how do the lesser throws (*adhareya*) 'go to the winning one (*kṛta*)'? Śaṅkara identifies the *kṛta* with number four, referring to the inferior ones as three, two, and one (*kali*). The lower numbers 'go to four', and are thus included (*antarbhavanti*) in the four. He envisions the relationship of the three losing throws with the winning one of four as that of 'numerical inclusion'. S. Radhakrishnan's translation follows Śaṅkara's interpretation: 'Even as all the lower throws of dice go to the winner with the highest throw, so whatever good men do, all goes to him.'[25]

[22]ChU 4.4.
[23]BU 2.4 and 4.5.
[24]ChU 7.
[25]S. Radhakrishnan (ed.), *The Principal Upaniṣads*, George Allen & Unwin Ltd., 1953, p. 401.

D. Shulman and D. Handelman describe games of dice in India[26] where the number four wins, offering a fine piece of existential interpretation:

In this perspective, there is an equivalence of sorts between the number that represents completion (four) and the empty hole; the latter suggests the reintegration of the fragmented holism which starts the game, when each hole contains an initial four seeds. To the extent that the board depicts the internal state of the player, one could say that the ideal winner is, paradoxically, full of emptiness—that is, of complete, fully reintegrated, and, therefore, empty holes.[27]

Thus, in this spirit, Raikva is a winner and whole hole (or 'full of emptiness'), while Jānaśruti is a loser, fragmentary and not whole. However, Jānaśruti's inferiority is complemented by a corresponding attraction—aspiring towards Raikva who is successful and whole.

We may further notice that from the very beginning of the story, Jānaśruti's somewhat inflated sense of confidence reveals some secret, underlying weakness. 'Everywhere, truly (*eva*) everywhere, people will eat my food' [*sarvata eva me 'nnam atsyanti*]. The Existential Reader may thus sum up the opening scene—one cannot avoid one's reality; since Jānaśruti has not found true peace (his economic success is somewhat deficient), the Upaniṣad is about to describe Jānaśruti's journey to Reality, consequent to his existential crisis. The existential crisis is about to be resolved by the transmission of knowledge.

The Traditional Reader is of course not totally blind to some of the existential dimensions in Jānaśruti's state of mind. In the conversation of the geese, the second goose's voice patently challenges Jānaśruti's self-esteem. Jānaśruti's 'inner crisis' in such a situation is quite obvious; the rest of the story lays bare Jānaśruti's misery and his quest for its alleviation. The Traditional Reader cannot overlook this dimension, and Śaṅkara says that the second goose's reference to Raikva's superiority was to Jānaśruti 'like an insult (contempt) to himself and praise for the other, a sage like

[26]Handelman and Shulman describe some Indian games of dice and offer existential interpretations in terms of emptiness, wholeness, etc. Thus, for example, they interpret the meaning of the number four in 'some variation on the very widespread Mancala series': 'Here we find a series of 14 empty pits or holes in which seeds or other tokens are deposited in turn, in accordance with rules that limit and regulate this "sowing"; distributing one's tokens, without having a remainder, is a major theme or goal. Units of four are critical: in the highly ritualized form of the game, a pit with three seeds in it can be completed if, and only if, the player has one seed left in his hand; in this case, the four seeds can be extracted as "winnings", leaving behind an empty hole.' (*God Inside Out: Siva's Game of Dice*, Oxford University Press, 1997, pp. 33–4).

[27]Ibid., p. 34.

Raikva' [*ātmanaḥ kutsarūpam anyasya viduṣo raikvādeḥ praśaṃsarūpam*].

However, the Traditional Reader does not attribute much significance and meaning to Jānaśruti's sense of inferiority, as the significance of the narrative lies not in the unfolding crisis but rather in the positive meaning of Jānaśruti's assets as a generous and humble householder. In his introduction to the story of Jānaśruti, Śaṅkara sees the narrative promoting 'easy comprehension' (*sukhāvabodha*) of knowledge and also the contribution of attributes such as faith, generosity, and humility as means of obtaining liberating knowledge (*śraddhāna-dānānuddhatatvādināṃ ca vidyā-prāpti-sādhanatvaṁ pradarśayata ākhyāyikāya*). Such a reading of the story ignores the Upanishadic narrative itself. According to the story in the *Chāndogya*, Jānaśruti's good feeling about himself—based on his generosity, etc.—is not conducive to a quest for knowledge. On the contrary, the quest for Raikva and his knowledge begins when all these praiseworthy traits of the householder become insufficient, as Jānaśruti awakens to his inferiority. Raikva, apparently a *sannyāsi*, is the opposite of Jānaśruti. Śaṅkara identifies his residence in the wilderness (*araṇya*), bringing him forth as an example of a renouncer who is thus superior to the one—like Jānaśruti—who lives in the village. Characteristically, the Traditional Reader extrapolates general attributes from Jānaśruti's individuality. While Jānaśruti's inferiority invites empathic reading, conducive to further identification with his personal suffering (seeking to establish a connection between his plight and the nature and content of the 'knowledge' he is about to receive), the Traditional Reader, faithful to Vedic otherness and resisting the dynamics of identification and empathy, focuses on impersonal traits. This Traditional Reader seems impelled to ignore the significance of the most fundamental part of the story, namely, Jānaśruti's crisis upon recognizing his inferiority. This fundamental evasion is characteristic of Traditional Reading of Upanishadic storytelling. Maitreyī's lost hope of immortality (or offspring), Śvetaketu's distress at being humiliated by his father, these features of the stories that invite empathic reading as an essential aspect of what this book exploratively calls Good-Enough Reading, are outside the ken of Traditional Reading approaches.

Towards the end of Jānaśruti's story, we hope he will be a winner and whole, with the transmission of knowledge as the resolution of inferiority. In the morning, Jānaśruti wakes up and says to his charioteer: 'Did you address me as if I were Raikva-of-the-cart?' For an Existential Reader—for whom Jānaśruti's psychological make-up is the crucially significant access to Jānaśruti's interior and crisis—this question exposes humiliation and neurotic pride rather than true humility.

Śaṅkara seems to be aware of Jānaśruti's predicament and sense of inferiority. He translates Jānaśruti's statement to his charioteer as follows: 'He (Raikva) is praiseworthy, but not I (*sa eva stuti-arho nāham*).'[28] And the charioteer—apparently ignorant about Raikva—answers with a question: 'Who indeed is this Raikva?' Jānaśruti responds by quoting the goose: 'As the inferior throws are contained in the *kṛta*, the winning four, so all good deeds are contained in this one (Raikva).' Jānaśruti thus accepts Raikva's superiority; he corroborates the goose's assessment. Jānaśruti's sense of security and his self-esteem—based upon his good (*sadhu*) deeds—are apparently injured. The charioteer goes away searching for Raikva, returns and says: 'I have not identified (found) him.' Jānaśruti sends the charioteer again, instructing him to search for Raikva where a brāhmaṇa should be. Śaṅkara says: 'In a solitary place, such as a forest.' Jānaśruti, himself a builder of villages and towns, is a paradigmatic householder, much involved in the world. If indeed he intimates wilderness (*araṇya*) as the place to search for a brāhmaṇa, the implication is that he concedes not only his own personal inferiority to Raikva, but that of the householder with respect to the renouncer. The charioteer goes away, and comes across a person seated under a cart, scratching himself. 'Are you Raikva-of-the-cart?' he asks. 'I am,' says Raikva. And the charioteer returns and announces: 'I have known (found, identified) him.' Jānaśruti then takes with him six hundred cows, a necklace, and a chariot drawn by two mules. Approaching Raikva, he announces: 'Raikva, here are six hundred heads of cattle, this necklace, and this chariot drawn by two mules. Please take this property, and teach me about your deity on which you meditate. Rejecting him, Raikva says: '*Śūdra*, let it (the necklace), together with the cows, be yours (*hāretva śūdra, tavaiva saha gobhir astviti*).' Jānaśruti returns home and takes one thousand cows, a necklace, a chariot drawn by two mules, and also his own daughter and goes back to Raikva. He says to Raikva: 'Here are a thousand cows, a necklace, a chariot drawn by two mules, this wife, and this village where you live. Please teach me.' Raikva maintains his allegation of Jānaśruti's inferiority; he addresses Jānaśruti as *śūdra*. However, heeding Jānaśruti's daughter's face, he says: '*Śūdra*, you have brought these; however, only by virtue of this face I will teach you (*ajahāremāḥ śūdrānenaiva mukhenālāpayīṣyathā iti*).' Thus, Raikva accepts Jānaśruti as a disciple, and teaches him.

Paradoxically, the story seems to end at the very moment when the

[28]Śaṅkara adds another possibility: 'Go to that Raikva-of-the-cart and tell him of my wish to see him.'

climax is forthcoming. For now, Śaṅkara would say, the real thing is about to happen (the transmission of teaching). However, Raikva's fascination with Jānaśruti's daughter's face and his acceptance of Jānaśruti as disciple constitute a—partial at least—closure of the narrative. The challenge suggested by the narrative lies in understanding Raikva's transformation in accepting Jānaśruti. How could a woman's face change the sage's heart? How could it overcome Raikva's resistance to teaching Jānaśruti? The tension of Jānaśruti's relentless quest for Raikva—sustained throughout the story—fades away; the satiated reader is not so eager to hear about the *ātman*. He ponders rather on the meaning of Raikva's openness to the woman's beauty; he may even imagine Raikva's future life with the rich man's daughter. The cohesive power of the unity-of-speech-act (*ekavākyatā*) is weakly maintained at this crucial boundary between the story and the *ātman*-message; thus Śaṅkara's theory of the Upanishadic story is not fully corroborated in this case. It is no accident that both Raikva and Jānaśruti disappear in the course of the teaching.

The story retold above contains some 'moments of interpretation'. We have suggested (observed) a correspondence between the geese's talk and Jānaśruti's inner voices; one voice corroborating the self-esteem of a householder, the other undermining the householder's pride. In Traditional Reading, such a correspondence—bordering on the existential—is not mentioned. In our Good-Enough Reading, Jānaśruti overhearing the geese's voices signifies the nature of his inner potential distress, opening him to a quest for 'knowledge'. Significantly, traditional commentary, while heeding Jānaśruti's distress, justifies the identification of Jānaśruti's distress through etymology rather than through 'experience'.

In his *Brahmasūtrabhāṣya*, Śaṅkara follows Bādarāyaṇa in his discussion of Jānaśruti's story in BS 1.3.34–35. The issue being raised concerns the *śūdra's* entitlement (*adhikāra*) to liberating instruction. Since Raikva addresses Jānaśruti as a *śūdra*, it seems—says Śaṅkara's virtual opponent—that the *Chāndogya Upaniṣad* is in favour of the *śūdra's* entitlement, since Raikva is seen teaching Jānaśruti. No, says, Śaṅkara; *śūdras* are not entitled (*ato na śūdrasyādhikāraḥ*). Jānaśruti is not born a *śūdra* (*itas ca na jāti-śūdra Jānaśrutiḥ*). So why does Raikva address Jānaśruti as if he were a *śūdra*? Śaṅkara answers in his Traditional Reader's voice:

The word 'śūdra' can be made to agree with the context in which it occurs in the following manner. When Jānaśruti Pautrāyaṇa heard himself spoken of with disrespect by the goose ('How can you speak of him, being what he is, as if he were like Raikva with the cart?' 4.1.3), grief (*śuc*) arose in his mind, and the rishi Raikva alludes to that grief with the word 'śūdra', in order to show thereby his knowledge of what is remote.[29]

If it be asked how the grief (*śuc*) which had arisen in Jānaśruti's mind can be referred to by means of the word 'śūdra', we reply: On account of the rushing in (*ādravaṇa*) of the grief. For we may etymologize the word 'śūdra' by dividing it into its parts, either as 'he rushed into grief' (*śucam abhidudrāva*) or as 'grief rushed in on him,' or as 'he in his grief rushed to Raikva', while on the other hand it is impossible to accept the word in its ordinary conventional sense. The circumstance (of the king actually being grieved) is moreover expressly touched upon in the legend.[30]

Śaṅkara's reference to the—somewhat far-fetched—etymology of *śūdra* while describing Jānaśruti's grief exemplifies the distance between his (Traditional) reading of the Upaniṣads and the Existential one. As Śaṅkara concedes, Jānaśruti's grief is obvious, and the story itself provides reason enough for the reader to notice it [*dṛśyate cāyam artho 'syām ākhyāyikāyām*). Pursuing Jānaśruti's inner being invites further elaboration of his crisis: He seems on an urgent quest for the resolution of a conflict. Does he have any idea about the forthcoming solution?

The narrative reaches its climax in the second encounter between Jānaśruti and Raikva; for Good-Enough Reading of the story it is crucial to perceive the identity of the heroes participating in the dialogue. Who is Jānaśruti? Who is Raikva? What is the meaning of Raikva's teaching *in the context of the narrative*? Raikva addresses Jānaśruti as a *śūdra*, but Śaṅkara vehemently denies such a possibility; hence W. Halbfass summarizes:

Śaṅkara's position is clear and, in its detail and rigour, goes far beyond the sutra text he is commenting upon. In his view, the śūdras may not be admitted to the study of the Vedas; they are to be excluded from the textual and educational access to the absolute unity of reality in the same way that (as the teachings of the *Pūrvamīmāṃsā* maintain) they are to be excluded from carrying out the Vedic ritual sacrifices. Śaṅkara presupposes that the *varṇa* system is based upon birth and physical family membership, and he makes it clear that the metaphysical unity of the real cannot in any way be taken as a premise of social and religious equality in an empirical sense.[31]

Śaṅkara discusses Jānaśruti's story in this context of the *śūdra's adhikāra*, and our interaction with the Jānaśruti's story may at times be more appropriate than Śaṅkara's. While in his insistence on Jānaśruti's identity

[29]See Thibaut's translation of Sāṅkara's *Brahmasūtrabhāṣya*, 1.3.34, p. 225.
[30]Ibid., pp. 225–6.
[31]'Homo Hierarchicus', in W. Halbfass, *Tradition and Reflection*, SUNY, 1990, p. 380.

(as a *kṣatriya*) he provides a valuable means of engaging the Upanishadic intention, in his preconceived idea of the independence of knowledge he may harbour a vested interest in favour of knowledge-based understanding, to the detriment of story relevance. An equation of the power of the sacrifice with the power of the story is essential to our understanding of Śaṅkara's theory of the Upanishadic story and his emphasis on the independence of knowledge (*svātantrya*). There is a built-in tension between Śaṅkara's assertion of the wholeness (*ekavākyatā*) of the Upanishadic story and his selective reading strategy, a tension which the Good-Enough Reader avoids through a consistent text-in-context grounding. Here a Good-Enough Reading must consider Jānaśruti's overt self-esteem as a householder, his lurking inferiority, its reflection in the geese's night-time conversation, his apparently inadequate approach to Raikva (assuming property to hold the key to Raikva's heart), Raikva's openness to Jānaśruti's daughter's beauty, skilfully woven together in a single narrative representing powerful themes. Suggested in the story is a hierarchy of values. Being a householder is not the apex of fulfilment. The renouncer is the great winner in life. As the inferior throws in the game of dice are included in the winning throw, so the householder's riches reach the renouncer. We may then interpret the story as expressive of man's (Jānaśruti's) yearning (and capacity) for self-fulfilment (or self-completion). This is probably a far-reaching interpretation. Yet why is Jānaśruti so disturbed by the geese's conversation? He could have brushed aside the second goose's reference to Raikva's superiority. Or he could have merely remained curious. But the story strongly suggests—as Śaṅkara notices—that Jānaśruti's response to the overheard avian conversation is a recognition of his own inferiority to Raikva.

There is no closure to the story. The Upaniṣad itself apparently dismisses the story, for we have nothing to read on Jānaśruti's fate. Does he return home? Does Raikva marry Jānaśruti's daughter? Is Jānaśruti significantly affected by Raikva's teaching? Does he overcome his sense of inferiority? Upanishadic indifference at this point is necessarily significant. The storytelling is powerful enough to create a remarkable impression of unfulfilment and of unfinished business. While some of the answers to the questions mentioned above (Does Raikva marry Jānaśruti's daughter?) may seem curious, strictly irrelevant to the 'contents' of Raikva's teaching, the same cannot be said of Jānaśruti's sense of inferiority, the driving force behind his quest for Raikva's knowledge. In this case, Good-Enough Reading begins with the recognition of the gap between the story and the message, a gap rarefied, rationalized, and congealed in the (Traditional and Existential) readings known to us. Interestingly, apparently beyond the boundaries of Jānaśruti's story, in ChU 4.3.8, the winning throw in the game of dice is mentioned again.

The 'theory of merger' (*saṃvargavidyā*) (ChU 4.3.1–4) is the knowledge transmitted by Raikva to Jānaśruti. Various elements disappear as it were, or are transformed, apparently into something more abstract than these elements. A single element, air (*vāyu*), is the 'merger' (*vāyur vāva saṃvargaḥ*). When fire is extinguished it vanishes into air (*yadā agnirudvayati vāyum evāpyeti*). When the sun goes down, it goes down into air (*yadā sūryastameti vāyum evāpyeti*). The moon sets into air. When waters dry up, they evaporate into air. Raikva refers to these cosmic mergers as divine (*adhidaivata*). Vital breathing (*prāṇa*) is similarly identified as a merger on the plane of selfhood (*athādhyātmaṃ prāṇo vāva saṃvargaḥ*). When one sleeps, one's voice goes to vital breathing (*sa yadā svapīti prāṇam eva vāg apyeti*). One's eyes, ears, and mind (*manas*) also go to that vital breathing. The Upaniṣad sums this up (ChU 4.3.4): 'These two are the mergers; air (*vāyu*) among the gods (*vāyureva deveṣu*), vital breathing among the vital functions (*prāṇaḥ prāṇeṣu*).'

To return to our story, the householder's self merges with a higher one. The transmission of knowledge shifts the momentum of Jānaśruti's existence from that of the householder, prompting the disappearance of inferiority. Can the articulation of knowledge—the *saṃvargavidyā* — cogently, convincingly, be considered as an event in Jānaśruti's life? What goes on in the course of Raikva's instruction to Jānaśruti? Raikva speaks to Jānaśruti. The storytelling seems to end. Abstract philosophy remains. Fire, sun, and moon disappear, extinguished into air. Waters are swallowed, disappearing into air. When one sleeps, speech merges with the vital breath, as do eyes, ears, and mind. How may Jānaśruti understand the *saṃvargavidyā*? How is such a theory of merger received? Such questions come to the fore in the course of Good-Enough Reading. While the answers to such questions may have been taken for granted in the distant past (say 800 BC) when the audience of Upanishadic speech was well versed in Vedic culture, having also experiences of a different order (those associated with meditation and renunciation), they are not readily available anymore. Renewed attempts to read Jānaśruti's story invite contemporary readers to try to see the possible meaning of the *saṃvargavidyā* within the narrative of the *Chāndogya Upaniṣad* 4.1.3. Jānaśruti's sense of inferiority originates in his inner split, abiding not in his householder role, but rather in the inner division itself, the unbridgeable gap between the householder's and the renouncer's existence. Jānaśruti is seen going to Raikva, to merge with him. The merger with Raikva (and the possibly concomitant renunciation) is a life-experience correlated with the theory of merger, apparently an abstract instruction in metaphysics. For Jānaśruti, such instruction may reflect his very own experience of merger leading to a new mode of wholeness.

The relationship of Jānaśruti's life with the *saṃvargavidyā* may be pursued further. The setting of the sun and the moon, the fire which is extinguished, the speech which is no more, the sight, hearing, and thinking which disappear in sleep, all are renounced as it were for the sake of a new consciousness, that of the air (*vāyu*) and vital breath (*prāṇa*). Raikva's instruction to Jānaśruti would imply the need for renouncing householdership in favour of celibacy and life in the wilderness. As the more abstract 'air' (*vāyu*) exists (or is made fully conscious) at sunset, and as the moon disappears and the fire is extinguished, so Jānaśruti's higher self—along with a complementary and consequent confidence—can materialize only on abjuration of the householder's self. As speech, sight, hearing, and thinking are necessarily absent in sleep, so must the inferior self disappear if the higher one is to come into being. The phases of renunciation and householdership are completely incompatible; the two are not possible at one and the same time. In other words, Raikva's *saṃvargavidyā* is addressed individually to Jānaśruti, who may be reluctant to forsake his householder's ground. Raikva points to the renouncer's self as better and more enduring, as *vāyu* and *prāṇa* are the more prevailing and real grounds compared with entities such as fire, sun, moon, speech, eyes, ears, and mind. In the light of his experience and awareness of inferiority, Jānaśruti is capable of understanding the *saṃvargavidyā*. In this vein, the Upanishadic storytelling concerns an individual's openness to deeply meaningful truths rooted in the life story of their recipients, as reflected in the way the philosophical message is embedded in the Upanishadic story. The very 'content' of the philosophical message is vitally connected to the storytelling. Jānaśruti's identification of the *saṃvargavidyā* as a message addressed to himself is thus crucial for our understanding of Upanishadic philosophy as 'forceful metaphysics'. As he overheard—by no accident—the geese talking about him, so too is he capable of recognizing the relational inferiority of the sun and the moon to air. Thus, the movement of the sun to *vāyu*, if commensurate with Jānaśruti's merger with Raikva, implies the sun's inferiority with respect to *vāyu* (Jānaśruti's inner split is thus also a cosmic one).

In this way, through such Good-Enough Reading, interpretation is essentially dependent upon a close reading of the narrative, which then must be read not as external to a philosophical message, but rather as an inherent part of it.

How have we read the story then? The Good-Enough Reading aspires to counteract the common phenomenon of dissociation of the story from the message, insisting on the essential hermeneutic integrity of narrative form and content. It views the Upanishadic speech-act as a unified act, insisting on the connection of story to philosophical contents, focusing especially on narrative voice and characters' speech in an assumption of contextual metaphysics. The Upanishadic assertion is dialogical, and the Upanishadic stories provide a framework of dialogues, inherently and saliently interactive. Yājñavalkya addresses Maitreyī; Uddālaka addresses Śvetaketu; Sanatkumāra addresses Nārada; Yama addresses Naciketas; the bull addresses Satyakāma; etc. This dialogic structure is exegetically significant. Its analytic exclusion in the interpretive strategies of the three reading traditions that I have distinguished from the one I call 'Good-Enough' is thus reductive, diminishing, and even distorting narrative content in their disregard for narrative form.

Good-Enough Reading cannot deliver a definitive reading of the Upaniṣads. It does, however, strive for a point on the continuum closer to complete, contextual reading, as I attempt to exemplify in the readings offered in the subsequent chapters of this book, thereby also tentatively theorizing Upanishadic reading. Śaṅkara's theory accounts for a mechanical mode of form/content relationship. Yet the underinterpreted gaps in story/message integration invite theoretical refinement, as I undertake to exemplify in this book in a twofold narrative of my own, which aspires to promote a theoretical conversation (on Upanishadic reading) and/through a close reading of several Upanishadic narratives.

Gaps remain, calling for greater comprehension. In other words, Good-Enough Reading is often not enough. It is an approach, an orientation often unfulfilled, engaging with a metaphysics of abstract yet powerful, charged ground, attracting particular beings and things to itself. Something happens to Jānaśruti as he listens to Raikva, absorbing his message. Humiliated, disconnected from his ordinary circumstances, in the foreign surroundings of Raikva's neighbourhood, Jānaśruti loses interest in and commitment to his householder's ontology and epistemology. An aspiration toward something of Jānaśruti's extirpated openness underlies all Good-Enough Reading. Its expression, as I hope to demonstrate, lies in its renewed attention to narrative form and content, to text in context, in reading that articulates both its aim and its awareness of the limitations of its label 'Good-Enough'.

TWO

Personal Crisis and Contextual Metaphysics: Reading the Under-Read Stories of Upakosala K. and Satyakāma J.

A reversal of values in reading the Upanishadic documents is suggested in this presentation of the Upanishadic experience, storytelling, and 'contextual metaphysics'. The crux of this reversal lies in the assertion that the knowledge transmitted in the Upaniṣads is revealed in the study of the narrative no less than that of its abstract contents. The narrative and the characters (Upanishadic heroes) have receded as the philosophy and mystical contents have gained prominence; thus the Upanishadic stories have not been read closely enough. A closer reading provides readers from different places and times a holistic mode of textual intelligibility. The early Upanishadic texts are viewed in this book as presentations depicting a movement from inferiority to excellence by force of 'knowledge', and the Upanishadic storytelling is assumed to be an indispensable map guiding us to the meaning of Vedāntic knowledge (or metaphysics). The articulation of the sublime metaphysics of the one omnipresent consciousness (*ātman*) is *an event*, an apparent happy resolution of the move from inferiority to excellence.

The transmission of knowledge, indeed, is invariably the end of the story. The Upanishadic narrative never goes beyond the event of knowledge articulation. The storytelling charts the transformation of the inferior to the higher self, to *ātman*-being. Thus knowledge in the early Upaniṣads serves as a transformational force guiding man and woman to a higher plane. The childless woman about to be deserted by her husband is bereft of immortality; she is capable of recognizing the 'knowledge' of non-dual reality as terribly necessary for her—a different road to immortality. The god Indra learns of inferiorities embedded in different states of the mind and thus becomes capable of knowing better.[1] Thus, a heightened sense of inferiority—as described in the Upanishadic narrative—breeds openness to 'knowledge.' Good-Enough Reading of the Upaniṣads begins with its narrative about inferiority and knowledge, a narrative which has the potential to transport its reader into the Upanishadic universe. As P. Ricoeur says: 'To understand, for a finite being, is to be transported into another life.'[2] One of the manifestations of less than Good-Enough Reading that this book attempts to investigate—and ameliorate—is the under-read text. Under-reading, as theorized in this chapter, is a less than full engagement with the textual source. More precisely, under-reading's partiality ignores the meaning potential of text-in-context. Extracting content for metaphysical speculation, under-reading thereby overlooks the significance of the story-as-told that a contextual reading can reconstruct. This chapter aims to exemplify the move towards contextual reading through a close reading of two stories.

Understanding requires inner space into which, so it seems, another (truth, for example) may enter. However, in ordinary life one is so full, hardly aware of any heaviness not nauseating enough to make one eager or hungry for space and openness. Reading the Upaniṣads is particularly frustrating since these texts require space which is mostly unavailable. We understand what we do in accordance with the scope of this—deplorably limited—space (or opening). Steeped in obstructive fullness, we understand what we understand.

The Upanishadic literature illustrates and focuses on the difference between life on the margin—characterized by insufficient inner space for understanding latent instructions about the self—and the adequate mode of receptivity to such instruction. Renunciation, collectedness, personal crisis, and awareness of suffering are among the circumstances conducive—even essential—to the creation of inner space commensurate with the emergence of the self (successful absorption of the teaching of the *ātman)*. The requirements for the creation of this inner space are fundamental to the mental culture which finds expression in the Upaniṣads. The inner conflict of the Hindu tradition, expressed as the tension between the householder and the ascetic, is a hallmark of the Upanishadic

[1]See ChU 8.7–12. Indra's journey through three modes of inferiority (wakefulness, dreaming, sleep without dreaming) ends in the excellence of the fourth state, timeless playfulness with women, relatives, chariots, etc. For a profound reference to the condition of play and playfulness, see D. Handelman and D. Shulman, *God Inside Out: Śiva's Game of Dice*, Oxford University Press, 1997.

[2]Ricoeur, *The Conflict of Interpretations*, Northwestern University Press, 1974, p. 5.

perspective. Even a truly great scholar such as A.B. Keith reveals a propensity to undervalue the meaning and role of non-verbal dimensions in the teaching of the Upaniṣads, significantly combining this tendency with doubt about the possibility of knowing the self:

> In these texts the vague requirements also occur that a man should be endowed with self-restraint, renunciation, tranquility, patience, and collectedness. But beyond these minor matters, which are none of them recognized as essential in the oldest Upaniṣads, a further question arises, how far there can be real knowledge at all of the Ātman.[3]

The Upaniṣads are preoccupied with means of creating inner space, and are replete with specifications and stories concerning the 'conditions of knowledge' (and the emergence of the self). Renunciation, patience, etc., are included among such conditions, and also—and this is a major theme in my essay—crises which embody the violent course of events invariably concomitant with the move from the margin to the centre.

Among the 'vague requirements' and 'minor matters', cited by Keith involved in the creation of efficient inner space, he does not include two central themes of the Upanishadic storytelling—crisis and the teacher (guru). These two aspects of the typical Upanishadic narrative are of course connected. As we shall see, throughout the stories of the Upaniṣads, the action of language is closely associated with the nature of guru/*śiṣya* communication. Personal crisis is often induced by the teacher himself, to instigate the creation of inner space necessary for understanding the self. Personal crisis is often described as a prerequisite for the march from the margin to the centre. Just before the last exposition (of nine repetitions of the 'same' saying) of the famous teaching (*Tat tvam asi*) to his son Śvetaketu, Uddālaka Āruṇi provides an illustration of this necessary association of crisis and the teacher. (As we shall see on a closer reading in Chapter 4, this is an opportunity to get a glimpse of Uddālaka's self-understanding as a teacher (and therapist)). In Āruṇi's story, a man born in Gāndhāra is cast blindfolded in a most solitary (and foreign) land. He shouts to the east, to the north, to the south, and to the west: 'I have been brought and cast here blindfolded.' This is a crisis. Śaṅkara hints that this is not just a fantasy. He says 'in the world' (*lok*), and elaborates on the evil nature and deed of the person who has done this; he is a robber (*taskara*) who also ties the poor man's hands and brings him to an

[3]A.B. Keith, *The Religion and Philosophy of the Veda and Upanishads*, Harvard University Press, 1925, rep. Motilal Banarsidass, 1989, p. 515.

unpopulated wilderness. Śaṅkara also provides more details of the poor man's condition; he is hungry and thirsty, the wilderness is infested with tigers and robbers, etc. The Upanishadic story reaches its climax: a good man approaches and instructs the blindfolded one on how to return home, and the latter does. From village to village he goes, until he is back in Gāndhāra, his homeland. The Upaniṣad concludes: 'A man who has a teacher—knows' (*ācāryavān puruṣo veda*).

Thus the *Chāndogya* tells a story of a possibly auspicious crisis and its remedy or resolution by a competent teacher. Śaṅkara's allegorical interpretation of the crisis refers to the nature of life in *saṁsāra* as an extended, chronic crisis; the poor man is snatched (by thieves who are one's good *and* bad deeds) from his true being (*sat*) which is the self of the world (*jagat-ātma-svarūpa*), and cast into the wilderness which is the body made of fire, water, food, air, bile, phlegm, blood, fat, flesh, bone, marrow, semen, worms, urine, etc.

While the role of the teacher (or Upanishadic therapist, guru) has been recognized as a constitutive component of the Upanishadic universe, the Upanishadic story itself has been subject to neglect. As a typical example of the hermeneutic potential of these under-read stories, we can note a single small implication of the story just discussed about the relationship of the famous Upanishadic pair—father and son—Uddālaka and Śvetaketu. (See Chapter 4 for an extended discussion on this story.) As mentioned above, the story of the blindfolded man's crisis is told by Uddālaka to his son in the midst of the famous *Tat tvam asi* discourse. What bearing does it have on our interpretation of the relationship of Uddālaka with his son? As everyone knows, the sixth chapter of the *Chāndogya* begins with Śvetaketu being sent away to study the Veda, and his subsequent homecoming as someone vain and conceited. The *Chāndogya* tells us that his father notices Śvetaketu's pride and initiates (precipitates) his son into crisis (by asking a paradoxical question). The commentators, scholars and translators—disregarding the story altogether—fail to raise questions about Uddālaka's state of mind. A son's pride and conceit may be bad; but is Uddālaka angry? Is he worried? Disappointed? Insulted? The issue is not altogether insignificant, since out of Uddālaka's state of mind emerges his *Tat tvam asi* teaching. Taking into consideration the larger context in which the story of the poor man from Gāndhāra is embedded, it seems likely that it reflects Uddālaka's self-understanding as a teacher who removes the blindfold from his son's eyes. And the state of mind—in this case—would be compassion.

The concept of understanding is the most apt concept of cognition with reference to the Upaniṣads. Understanding is less tangible than knowledge or insight. Indeed, in everyday life understanding involves a process of widening of ego boundaries and implies comfort, ease, light, vibration, excitement, pleasure, connectedness. The Upanishadic authorities maintain that sometimes understanding is immensely rewarding and associated with relief, immortality, liberation, self. This is the reason why it is so desperately sought. But even as we utter 'immortality' and 'self' we feel a little pain, a shadow of constriction, incomprehension. We do not have the space necessary for such ideas, 'immortality' or 'self'; we have not understood.

The crisis stories, together with other requirements (such as recollectedness and renunciation), are part and parcel of the unified speech-act of the Upaniṣads. And of course the Upaniṣads *do* contain various formulations of the latent teaching of the self. These are manifested, however, in occurrences of heated language inaccessible to the margin. But as Keith suggests, understanding the self (*ātman*) is most difficult; it requires crisis and discipline and painful transit from the margin to the centre.

Apparently, Keith has in mind Upanishadic statements such as in the *Taittirīya Upaniṣad* 2.4.1[4], or in episodes such as Yājñavalkya's reply to the Vedic scholar Uṣasta Cākrāyaṇa, in the BU 3.4.2. 'You cannot understand the understander of the understander' (*na vijñāter vijñātaraṃ vijanīyaḥ*). This latter episode is revealing, since it is a very concise story which exemplifies the distinction between (heated, contextual) instruction and (cold, context-free) metaphysics.

The Upaniṣad describes Uṣasta's crisis at not understanding the *ātman*; stuck on the surface, Uṣasta is exasperated by Yājñavalkya's preliminary description of the self as what exists within all and within his own (Uṣasta's) being, and also as the source of breathing, the agent which 'breathes the breathing', etc. The scholar, facing the sage, confesses incomprehension of the *ātman* and wishes—demands, actually—a more direct (*sākṣāt-karaṇāt*) and satisfactory delineation of the self. The sage, Yājñavalkya, responds by asserting that 'you cannot understand the understander of the understander,' and similar statements, such as 'you cannot hear the hearer of hearing,' etc. But these assertions—mysteriously—plunge the listener into the furthest recess of subjectivity and selfhood, thereby producing a moment of understanding. The great paradox here (BU 3.4.2) is that the statement of the unknowability of the self does elicit a moment of

[4]'That from which speech withdraws ...' (*yato vaco nivartante. ...*)

understanding (or knowledge). Uṣasta, indeed, understands and becomes silent. In contrast, from his position on the margin, Keith interprets Upanishadic language as strictly referential, thereby reaching his interpretation of the unknowability of the self (according to Yājñavalkya). Yet a return to the Upanishadic narrative suggests that there is more to be won from this story. Yājñavalkya's apparent denial of knowledge of the self precipitates precisely what it seems to deny on a strange shore of seemingly impossible understanding. Reference is not the only action of language, and Upanishadic episodes such as this one clearly demonstrate the limitations of the marginal perspective.

The distinction between the margin and the centre is often emphasized in the Upaniṣads themselves. Existential crisis ('sorrow') is explicitly connected with the crisis of margin language. Such language-in-crisis is labelled 'mere names' in the instructive story of Nārada and Sanatkumāra in the *Chāndogya Upaniṣad* 7.1. This story is explicit in its connections between life on the margin, the crisis of language and personal pain. In this story, the Vedic scholar Nārada asks Sanatkumāra to teach him. The latter asks Nārada to tell him what he (Nārada) already knows (*yad vettha*) so that he (Sanatkumāra) could teach him something higher (*ūrdhvam*). Nārada seems densely full of knowledge, and he responds with a long (somewhat grotesque) list of seventeen (!) fields of knowledge (or types of texts) which he has read and mastered; these include the four Vedas, mythology and history, mathematics (*rāśi*), grammar (the Veda of Vedas, *vedānāṃ vedaḥ*), science of astronomy, astrology, political science, science of spirits, serpent wisdom, etc. Nārada's long list encompasses and exhausts the language of the margin, and Sanatkumāra refers to all the manifestations of this language (including the four Vedas) as 'mere names' (*nāmaivaitat*). Nārada's own self-understanding corresponds in this respect with Sanatkumāra's implicit analysis of language. And Nārada complains about being merely a knower of verbal texts (*mantra-vid evāsmi*), not a knower of the self (*ātma-vid*). Śaṅkara forcefully identifies the crisis of margin language with its merely referential power, and paraphrases Nārada's self-understanding thus: 'Though I know everything (*etat sarvaṃ jānannapi*) I am only a verbalizer (*mantravid*) in the sense that I know only the meanings of words (*śabdārtha-mātra-vijñānavān asmi*), namely, the names (or references) (*abhidhāna*) of the words (*sarvo hi śabdo 'bhidhāna-mātram abhidhānaṃ ca sarvaṃ mantreṣvantarbhavati*').[5]

The Upanishadic storytelling focuses here on the painful existence

[5]See ChUBh 7.1.3.

on the margin. Thus Nārada expresses his own personal crisis when he asks Sanatkumāra to take him 'beyond crisis'. For indeed, *śoka* (sorrow) is not a temporary state of affairs in this context, but a condition of chronic crisis. The emergence of the self ('knowledge of the *ātman*') (*ātmavidyā*) is a condition contrasted with objective knowledge (*mantravidyā*). Thus Nārada has heard that whoever knows the *ātman* crosses beyond the condition of sorrow; he also confesses to his painful existence (*so 'haṃ bhagavaḥ śocāmi*) and asks Sanatkumāra to help him cross to the other shore of sorrow (*taṃ mā bhagavāñ cchokasya pāraṃ tārayatu*). The recognition of nauseating fullness leads and amounts to a confession comprising warmer language, closer to the hot centre (where understanding delivers from sorrow). Most instructive in this context is the Upanishadic prescription (for Nārada only!) of the movement from the margin to the centre in terms of understanding the hierarchy of speech-acts. Is there anything higher than the speech-act of mere naming? Yes, my dear, 'speech' (*vāc*) is higher than mere naming (*vāg vāva nāmno bhūyasi*). Is there anything higher than speech? Yes, mind (*manas*) is greater than speech. And 'will' (*saṅkalpa*) is higher than mind, 'intelligence' (*citta*) above will, meditation (*dhyāna*) higher than intelligence, understanding (*vijñāna*) higher than *dhyāna*, 'power' (*bala*) above understanding, food (*anna*) greater than power, water (*āpaḥ*) greater than food, fire (*tejas*) greater than water and space (*ākāśa*) above fire. What is higher than space? Memory (*smara*) is higher than space and 'hope' (*āśā*) greater than memory. The vital force (*prāṇa*) is higher than hope, and the hierarchy continues to unravel.

Language has temperature. On the margins it is cold; at its centre it is heated. The Upaniṣads demonstrate how variable language is in terms of temperature and action. At the core there is full-body, transformative, hot understanding and transmission. There, in the dangerous interior, language makes one effulgent: 'Your head glows like that of one who knows the Absolute,' says the teacher, observing the disciple's appearance[6] after the latter's encounter with fiery language. At the very centre, language is most present and real. There, in reality, it transforms Śvetaketu, as his father asserts: *Tat tvam asi*. On the margins—where we reside—language and understanding are different. From the outside we watch others' excitement, we behold heroes in crises, ripening in their exposure to fiery language; staying on the margin, we remain their shadows. The more remote from the centre, the more abstract and irrelevant language becomes, even dwindling into the condition of 'mere words.'

[6]See ChU 4.14.2.

Since in the Upanishadic literature, some 'cognitive events', comprising language, occur at the centre, there is some justification for the concept of 'metaphysics' applied to these events. However, a temperature-sensitive description of the Upanishadic language (text) must also take into account the theory of language and language-heat implicit in the Upaniṣads, a theory focused on the hierarchy of speech-acts in terms of their relative distance from the core. Since this sensitivity to heat is a major focus of attention and interest for the Upanishadic speaker, the context of cognitive events (occurrences of partial or complete awakenings, etc.) is relevant. Thus—and this is our primary presupposition—storytelling in the Upaniṣads reflects and redacts a theory of language. With language-heat (temperature) being so important, questions arise about the circumstances, causes and contingencies of particular sayings (say, *Tat tvam asi*). We—on the margin—decode such a statement in language like this: 'This famous text emphasizes the divine nature of the human soul, the need to discriminate between the essential self and the accidents with which it is confused and the fetters by which it is bound.'[7] Such deracinated interpretations grow in decontextualized soil; reintegration of text in context reveals textures and meanings lost as philosophical texts are extirpated from narrative contexts. Note the difference between such a statement and Uddālaka's utterance *Tat tvam asi, śvetaketo*. The context of the *Chāndogya Tat tvam asi* is the crucible where the afflicted son implores and beseeches his father: 'Father, teach me more!' (*bhūya eva mā bhagavān vijñāpayatv iti*). Uddālaka's teaching is not classroom metaphysics but a father's response to the urgency of his son's call. Moreover, in Uddālaka's mouth this 'statement' recurs nine times. Are all the speech-acts the same (surface identical but contextually different)? Not necessarily. For if each speech-act is characterized by a particular temperature, the first *Tat tvam asi* may be colder than the second, and the final may be the hottest, as Śvetaketu's consciousness merges with his father's or with the Absolute. (Thus, in the course of his dialogue with his father, Śvetaketu heats until he boils in understanding).[8]

In my view, the Upanishadic theory of language (or language heat) is expressed in and through storytelling. The stories contextualize crucial manifestations of language. An individual story may be (and has been) construed as the 'mere occasion' for transmission of (context-free)

[7]See S. Radhakrishnan (ed.), *The Principal Upaniṣads*, George Allen & Unwin Ltd., 1953, p. 458.

[8]See Chapter 4 for further discussion of this much read under-read narrative.

metaphysics, or alternatively as a guide to and means of interpretation of Upanishadic speech-acts. Following this second approach, the more important the story, the more significant the context of speech. Taking into account the story meaningfully is contextualizing metaphysics, bringing it closer to its centre, where it is hotly significant. In contrast, out of context, metaphysics is 'universal', as understood on the margin, from the outside.

Contextual metaphysics is difficult and somewhat inaccessible, yet the Upanishadic stories invariably point in this direction. As we will see, personal crises such as death,[9] rejection by one's teacher,[10] potentially lethal uncertainty over one's father's identity,[11] the emergence of criminal desire,[12] challenging one's mother,[13] separation from a spouse,[14] etc., are all necessary and fertile grounds for healing Upanishadic instruction.

While life on the margin has its comforts, it also has some apparent disadvantages. The primary disadvantage, in this view, is the difficulty in experiencing, learning, or understanding the kind of contextual metaphysics explicated in the Upaniṣads. In order to understand what we understand one must translate the Upanishadic inner space into our colder marginal language. We are tempted to translate heat-laden language into context-free metaphysics, in spite of the fact that contextual metaphysics is powerful and alive, whereas context-free metaphysics is dull and much weaker. Yet if the story is given due importance, its context may nurture our metaphysical understanding. A study of the narrative may thus facilitate more sensitive reading of even the most abstract and universal expressions of the Upanishadic voice. Thus we can hear the big difference between the heated, Upanishadic *Tat tvam asi* addressed *at a particular moment by father to son* and the *Tat tvam asi* in, say, Śaṅkara's numerous references in the *Brahmasūtrabhāṣya* and elsewhere. Śaṅkara's citations of *Tat tvam asi* are complex, of course; much warmer than my references to them and colder than Uddālaka's.

However appealing the concept of 'contextual metaphysics' may be, it presents an essential difficulty here. The Upanishadic story might be interesting and instructive in the limited 'context of discovery', but irrelevant with respect to the justification or interpretation of the

[9]See for example Naciketas' story in KU 1.
[10]See Upakosala's story in ChU 4.10.
[11]See Satyakāma's story, in ChU 4.4.
[12]See Upakosala's story.
[13]See Satyakāma's story.
[14]See Maitreyī's and Yājñavalkya's story in BU 2.4.5.

Upanishadic teaching. Let us suppose Yājñavalkya tells his wife Maitreyī that he is going to the forest, leaving her behind, and also that she is indeed stunned, overwhelmed by the prospective separation from her beloved husband, who is about to go to the forest of his own free will (no external circumstances are mentioned). Then let us suppose Maitreyī is in crisis and seeks her husband's teaching so that she may become 'immortal,' and also that Maitreyī's personal crisis makes room (inner space) for her receptivity to the tacit teaching of the *ātman*, which Yājñavalkya does impart to her. All this is the context of Maitreyī's openness to instruction. But how does it affect the *meaning* of the instruction or our interpretation of the metaphysics of the self? Conventionally, the meaning and interpretation of the respective metaphysics is essentially *independent* of the occasion of its transmission. Let us suppose the apple did fall on Isaac Newton's head and thus trigger the theory of gravity. We can then understand and justify the theory of gravity without the apple and Newton's head. Is the context of Yājñavalkya's teaching to his wife-in-crisis different? The truth of Yājñavalkya's theory of the self is judged not by its efficacy for a woman named Maitreyī, but by virtue of other criteria. Who really cares about Nārada's condition of despair, his existential condition of *mantravid*? Sanatkumāra's teaching of the origination and hierarchy of language instances from brahman to mere names can and should be validated on grounds such as truth, coherence, compatibility with evidence, etc. Nārada's, Śvetaketu's, Maitreyī's, Upakosala's, Indra's, Uddālaka's, Satyakāma's, Naciketas', and Yājñavalkya's crises are (conventionally) different contexts of discovery, essentially irrelevant to justification and interpretation of the theory of self (or language) discussed in the Upaniṣads.

This is a powerful objection, which touches the core of the discourse concerning the value and meaning of the Upanishadic story. Moreover, the Upanishadic teaching of the self seems to 'recur' in different contexts; why assume then that context is so important? Moreover, the history of interpretation of the Upaniṣads has taken the context-free nature of the Upanishadic teaching for granted. Yet the Upanishadic language deals with the self, and the basic referential instinct on the margin obstructs understanding of the Upanishadic text. Utter helplessness is thus the marginal experience of the interpretation of the Upanishadic teaching. How then shall we understand Satyakāma's final instruction to Upakosala, his disciple? He says: 'The person seen in the eye is the self, *ātman*, fearless and immortal.' A referential axis is of no assistance to us.

Most scholars and commentators exist and think on the margin, at a safe—but useful—distance from the centre. The greatest interpreters and translators such as Śaṅkara, Vācaspati Miśra, and Ānandagiri, along with Keith, Deussen, Oldenberg, Ranade, Radhakrishnan, Gambhirananda, Olivelle and Müller, make constructive use of their position on the margin. However, they are—each in his own way—also full with their own knowledge, attitudes, and ideas. Fullness may have its advantages and uses, but it is not everything. When Deussen compares the Upaniṣads with the New Testament, claiming that 'the Bible sees corruption in the willing part of man, the Upaniṣads see it in the knowing part of man; the Bible promotes the transformation of the will, the Upaniṣads the transformation in knowledge',[15] he reveals a particular mode of fullness, both useful and obstructive. Similarly, Deussen's association of Moses and the burning bush in his presentation of Satyakāma's story is also a token of fullness on the margin.[16] Numerous indeed are the symptoms of fullness and spiritual satiety.[17] Indeed, the Upanishadic heroes—Śvetaketu, Upakosala, Indra and Virocana, Jānaśruti, Uddālaka, Satyakāma, etc.—struggle with the same predicament of premature and preemptive fullness, like all of us. The renunciation and substitution of available being for unavailable (unknown) existence is a primary theme here. Explorations of the Upanishadic story map the hazards and costs of fullness.

Thus, let us assume that we are exiled and distant from the space where the Upanishadic teaching takes place, and that we really do not understand. Let us further explain our frustration by the argument that the Upanishadic metaphysics is essentially *contextual* rather than abstract, and that the Upanishadic story is the location of this contextuality. Thus, when Uddālaka Āruṇi utters the most famous saying of the Upaniṣads—*Tat tvam asi* ('you are that')—he directs his message to his son Śvetaketu as the latter becomes free of the conceit and self-aggrandizement that characterized him at the beginning of the story.[18] The connection between Śvetaketu's pride when he returns home and the 'metaphysical message' known as *Tat tvam asi* provides the great *mahāvākya* with an essential contextuality. Indeed, the strong indexical language here (*Tat tvam asi*) is so pronounced that much effort is needed to extirpate such heavy contextuality!

[15]P. Deussen, *Sixty Upaniṣads of the Veda*, trans. by V.M. Bedekar and J.B. Palsule, Motilal Banarsidass, rep. 1991, p. xii.

[16]Ibid., p. 122. Deussen compares Satyakāma's attendance on his guru's cows and his contact in the wilderness with Moses' in the desert.

[17]See discussion of the margin in Ch. 2.

[18]See Ch. 4, the discussion of ChU 6.

However, it is also true that such messages are easily abstracted from the narrative and made into context-free 'philosophy'. The dominant mood in the reception of the Upaniṣads has been—most obviously—the abstraction of the metaphysical message—and concomitantly—the bifurcation of the Upanishadic text into 'metaphysics' and 'story'. I aim to do the opposite; my direction is thus contrarily restorative.

One of the more significant aspects of the Upanishadic storytelling which is also salient to contextual hermeneutics is the individuality of the Upanishadic spiritual heroes. The long tradition of commentary and scholarship on the Upaniṣads has relegated the character and personal crises of teachers, disciples, wives and mothers, sons and gods, to an irrelevant background (as is evident in Oldenberg's view of them as 'clumsy sketches' of everyday life).[19] Even the most famous of the Upanishadic heroes have not attracted attention in their individual life stories. Take for example Śvetaketu's pride and vulnerability,[20] his father's anxiety[21] and humility,[22] Yājñavalkya's courage and assertiveness,[23] Upakosala's depression, Indra's doubts,[24] Virocana's complacency, Maitreyī's revolt and fear of loneliness,[25] Nārada's despair of scholarship,[26] etc. Salvaging of these heroes' individuality and experiences might contribute to a better understanding of our location in relation to the text and to remedying our incomprehension of the emergence of self according to the Upaniṣads. In general, the ripe and deep thinkers of the Upaniṣads have been doomed to oblivion in the Western horizon. See for example, the deepest and most comprehensive discussion on the mutual perception of India and the West. In Halbfass' *India and Europe* there are references to Western spiritual heroes such as Moses (six references) and Jesus (thirteen references), but no reference at all to the great Upanishadic thinkers such as Uddālaka Āruṇi or Yājñavalkya. While the Upanishadic texts do occupy a prominent place in the Western consciousness of India, the Upanishadic protagonists and storytelling recede to a negligible background.

But is the Upanishadic storytelling sufficiently complex and interesting

[19]See discussion on p. 55.

[20]Śvetaketu's vulnerability is beautifully captured in the story of ChU 5.3.1–5.

[21]See for example Uddālaka's refusal to teach the five rich householders, according to ChU 5.11.4.

[22]See the story of his becoming king Jaivali's disciple, in the ChU 5.3.6–9.

[23]See for example BU 3.

[24]See ChU 8.7.

[25]See BU 2.4 and 4.5.

[26]See ChU 7.1

as a relevant context for the Upanishadic teaching of the self? Do the protagonists possess true individuality? Before one can consider the connection of context with metaphysics, it is necessary to offer a reading of stories as a mapping of the potential for hermeneutic reciprocity. It is a waste of time to address Upanishadic storytelling with such expectations if it consists of no more than descriptions of Socratic seers or philosophers seated either on river banks or amidst grazing cows. But if the Upanishadic stories offer sufficiently complex characterization, there may be some value in them. My first goal is merely to explore the narrative potential of such stories—in terms of narrative complexity, characterization, conflict and development, etc.—to incubate metaphysical teaching. Later on (Ch. 4) I shall also make a few suggestions with regard to the meaning and force of the Upanishadic story with respect to the interpretation of the 'Upanishadic message'.

The following is a discussion of two stories told in sequence in the fourth section of the *Chāndogya Upaniṣad.* The heroes are Satyakāma J. and Upakosala K., two young men who face crisis and gain access to new modes of inner space. In the course of their crisis and agony they discharge (renounce) enclaves of fullness and thus obtain auspicious emptiness (into which spiritual instruction may enter). In the course of their progressive renunciation these two men touch and taste language in its heated, close-to-the-centre, transformative modes. Upakosala is rejected by his guru, renounces his (unlawful) desires, receives the lukewarm (yet far from cold) teaching offered to him by an alternative source (the three fires), and then renounces his lawful (dharmic) desires and re-unites (with fear and trepidation) with his teacher. The communication with the teacher provides the context for the action of the hottest (non-referential) language. Satyakāma seeks to know his father's identity, denounces and forsakes his *śūdra* mother, is accepted and rejected by his newly found guru, receives instruction from alternative sources, and re-unites (with fear and trepidation) with his teacher. The instruction—'metaphysical message'—is possibly related to the individual nature of Satyakāma and Upakosala in their quest. Indeed, the harsh transcendence of normal identity implied by the Upanishadic message makes a complementary story of crisis, rejection and despair into a natural completion of a unified speech-act.

At a crucial moment in his life, Upakosala K. is cast alone, rejected by his teacher (guru). At the apex of his training, ready to become a *snātaka*,[27]

[27] A *snātaka* is one who took the ritual bath at the end of his *brahmacārya* condition. See R. Lingat, *The Classical Law of India*, University of California Press, 1973, p. 46.

leave for home and participate as a householder in the maintenance of the fires, family, and the universe,[28] Upakosala is denied instruction. According to the story told in the *Chāndogya Upaniṣad* 4.10.1, of the students residing at his guru's home, he alone is left over, dangling as it were, uninstructed; the *Chāndogya*, which provides very concise, telegraphic story contexts, is explicit on this point: though the teacher initiates all the other disciples, he does not initiate (and does not dismiss) Upakosala (*sa ha smānyān antevāsinaḥ samāvartayaṃs taṃ ha smaiva na samāvartayati*). The teacher's wife, a seemingly soft feminine presence in the predominantly male atmosphere of the guru's house, is distressed at the sight of Upakosala's singular neglect and sorrow. Approaching her husband, she protests: 'This disciple is heated (tormented) (*tapta*), and (he always) tended the fires well (*tapto brahmacārī kuśalam agnin paricacārīt*).' In spite of her subordination to and dependence on the goodwill of her husband, a note of warning is woven into the teacher's wife's appeal: 'Before the fires overtake you, teach him!' (*mā tvāgnayaḥ paripravocan prabrūhī asmā iti*). But the teacher remains silent and leaves home without teaching Upakosala. What is going to happen now? The young man is left behind by his teacher; not alone, to be sure; he lives in his master's house with his master's wife. Diseased and afflicted, he is resolved to abstain from food (*sa ha vyādhinānaśitum dadhre*). The teacher's wife inquires: 'Why don't you eat?'

We may wonder at this point about Upakosala's condition; apparently, he has good reason to feel dejected, having been rejected by his teacher. Is not being rejected by one's 'spiritual father'[29] reason enough for being sick unto death and refusing food? But does not the teacher's wife know this? And if she does, why does she ask? What then is the meaning of Upakosala's refusal to eat? We must turn to Upakosala's own answer to the teacher's wife's question (why don't you eat?), i.e. to Upakosala's self-understanding of his crisis. He says: 'Many and of many kinds are the desires in this man; I am filled with afflictions. I will not eat.' (*bahava ime 'smin puruṣe kāmā nānātyāyā vyādibhiḥ pratipūrṇo 'smi nāśiṣyāmīti*). Thus, Upakosala refers to his condition of *multiple desires* as the contents of his fullness and as the reason for his refusal of food. He accepts responsibility for his condition, and in a sense justifies his teacher's decision. Indeed,

[28]About the role of the householder in the maintenance of the universe (by sacrifices, etc.,) see B.K. Smith, *Reflections on Resemblance, Ritual, and Religion*, 1986.

[29]See *The Laws of Manu*, trans. by W. Doniger and B.K. Smith, Penguin Books, 1991, p. 2.

much silence envelops Upakosala's person. His reflection on the condition of multiple desires is Upakosala's first speech-act in the story; he does not argue or complain about his rejection by his brutal teacher. He seems rather to contemplate his weakness.

If we think of Upakosala's predicament as a token of inferiority (in comparison with the other students, who have been sent home) or as inexplicably tough luck, we may be wrong; for the narrative suggests that Upakosala is *chosen* rather than victimized. The other students have been discharged homeward, to become householders. The Upanishadic mental culture is explicit on the vast difference between renunciation—which leads to the way of the gods[30]—and the way of the fathers,[31] the householder's lot. Indeed, Upakosala is launched on the road to further heating. His confession of multiple desires is a heated speech-act, a mode of *tapas* similar to his fast.[32] The 'Upanishadic speaker' narrating the story probably approves of the teacher's wisdom in this context. The teacher seems to have diagnosed Upakosala's state of craving and made a move to bring the point home to the disciple. Thus, the teacher's leaving—a crisis of rejection—is followed by the disciple's openness to his condition of being 'full of desires'. The second crisis is impossible without the first one.

Getting closer to the story, we must ask: what is the nature of Upakosala's desires? (In order to approach the contextual core one needs to explicate the context.) Śaṅkara conceives of Upakosala's predicament as consisting of 'unfulfilled wishes'. He renders 'desire' (*kāma*) as 'wish' (*icchā*) and describes Upakosala as one judgemental of himself as a common man (*prakṛta*), whose goals—or wishes—are not fulfilled (*akṛtārtha*). And what are Upakosala's wishes according to Śaṅkara? He conceives of Upakosala's troubles as mental agonies (*cittaduḥkha*) rooted in his incapacity to carry out what should be done (*kartavyatāprāptinimittāni cittaduḥkhāni*). Śaṅkara

[30]See ChU 5.10 and BU 6.2.

[31]The way of the gods (*devayāna*) is the road from which there is no return. The way of the fathers (*pitṛyāna*) leads to the moon, where the residue of karma is exhausted, and through the semi-natural cycle of rain one is absorbed back on earth, in a precarious condition of waiting to be born again through being eaten by some animal or man. See discussion in W. Halbfass, 'Competing Causalities: Karma, Vedic Rituals, and the Natural World,' in W. Halbfass, *Tradition and Reflection*, SUNY, 1990, pp. 291–346.

[32]Fast is a well-known channel of 'austerity' (*tapas*). However, as Handelman and Shulman emphasize, *tapas* is a creative process: '... *tapas* tends to trigger an upward transition in levels of being or awareness—and this transition is naturally accompanied by the anguish inherent to growth.' (*God Inside Out: Śiva's Game of Dice*, Oxford University Press, 1997, p. 162).

is possibly aware of a rule such as *Manusmṛti* 11.204: 'If a man neglects the daily rituals prescribed by the Veda or fails to fulfil the vows of a Vedic graduate, fasting is the restoration.'[33] Thus Śaṅkara thinks of Upakosala as a good boy, possibly a perfectionist, who is dissatisfied with himself. But why is Upakośala so keenly aware of his shortcomings after having been rejected by his guru? Śaṅkara's interpretation does have real logic here. The student seeks (and finds) reasons for his rejection. And yet, Śaṅkara's interpretation is not perfect, since Upakosala is described by the teacher's wife as worthy of instruction, and Upakosala mentions no unfulfilled wishes, etc., simply saying he has many desires of many kinds. Moreover, Śaṅkara diagnoses Upakosala's depression as similar to Nārada's sorrow. But Upakosala and Nārada differ in their state of mind and depression. Upakosala is much younger, and the nature of his fullness is necessarily different from Nārada's. He is full of desires, not of surface-knowledge. The language of *kāma* combined with the means of repentance—fasting—suggests *unlawful* desires rather than lawful ones.

In the course of his crisis, Upakosala K. moves—unaware perhaps—toward a larger inner space, which is openness to instruction. It seems that Upakosala's confession to his master's wife does bear some fruit. He becomes open, and the wife's admonition to her husband comes true, and the fires approach Upakosala and teach him. Upakosala has apparently undergone a change of heart, a transformation which makes him ripe and receptive.

Let us focus again on the nature of Upakosala's change of heart. What desires does Upakosala find in himself? Let us look at the bare facts again: the student (Upakosala) and the guru's wife share their home as the guru leaves. The student is heated, afflicted, and does not eat. The woman cares for the young student and wishes to feed him. The student confesses intimately to the guru's wife: 'I am full of desires.' Though the Upaniṣad does not explicate Upakosala's desires as sexual ones, focused on the guru's wife, the story is strongly suggestive of such a possibility, which is, I think, somewhat more likely than Śaṅkara's. The active presence of the teacher's wife, the absence of the teacher, the woman's wish to feed Upakosala, the latter's inner heat (accumulated through years of continence) and confession of multiple desires point in this direction.

Upakosala confesses to the guru's wife a matter concerning one of the

[33] *The Laws of Manu*, trans. by W. Doniger with B.K. Smith, Penguin Books, 1991, p. 270.

gravest sins of Vedic India: the defilement of the guru's bed. Punishments for this crime are incredibly harsh, according to the law books.[34] The Upaniṣad does not say that Upakosala sleeps with his preceptor's wife, but suggests the respective desire in Upakosala's (and maybe also in the wife's) mind. The teacher's absence from home creates an opportunity for Upakosala to realize his condition of desire and to transcend it. Upakosala's confession could be the renunciation of unlawful, adharmic desire, so that a new opening and fresh inner space become available. His teachers at this moment of his life are the three fires of his guru's home, the fires expressive of his harsh mentor's selfhood.[35] In fact, 'installed at the time of marriage it (fire) defines the household as a unit. ...'[36] In this sense also, in attending to the fires, Upakosala replaces the absent teacher.

The fires share the mood of the guru's sensitive wife; they consult among themselves and decide: the student here is heated and has taken good care of us. Let us teach him! And the fires, collectively, impart to Upakosala the teaching: *prāṇo brahma kaṃ brahma khaṃ brahmeti*. This is then the truth commensurate with Upakosala's newly acquired inner space. What does it mean? Do the fires simply utter these three equations (Brahman is *prāṇa* (breath). Brahman is *ka* (joy). Brahman is *kha* (space))? What does it mean to understand such 'propositions'? Do we—readers of the Upaniṣad, cast away on the margin—understand *prāṇo brahma*? What precisely is *prāṇa* (life force, breathing, subtle energy, etc.)? What is brahman? The text is extremely ambiguous. Perhaps as the fires utter their message it is heated, contextual, and transformative. Doomed to marginality and context-free metaphysics, we do not understand it.

Upakosala could hear such utterances before, but his hearing did not amount to instruction, his hearing did not produce the comfort of understanding. Following the narrative, we are faced with a mini crisis by now. We do not understand *prāṇo brahma*. Upakosala is somewhat better off. He says he does understand *prāṇo brahma* but he does not understand *ka* and *kha* (*vijānāmi ahaṃ yat prāṇo brahma kaṃ ca tu khaṃ ca na vijānāmīti*). Upakosala does not comprehend *ka* and *kha*. P. Olivelle translates and completes the meaning: 'I can understand that brahman is breath. But I do not understand how it can be joy or space.' (p. 133). The fires answer:

[34]See *Manusmṛti* 11.104–107.

[35]See D. Knipe, *In the Image of Fire*, Motilal Banarsidass, 1975.

[36]See J.C. Heesterman, *The Broken World of Sacrifice*, University of Chicago Press, 1993, p. 86.

yad vāva kaṃ tad eva khaṃ yad eva khaṃ tad eva kamiti. Prāṇaṃ ca hāsmai tad ākāśaṃ cocuḥ.

What was Upakosala's difficulty? Could he not experience brahman as *ka* and *kha*? But then what is *ka* and *kha*? Less fortunate, we cannot even experience Upakosala's difficulty. Thus, unlike us, Upakosala seems successful, understanding the fires' explication.

Let us continue with the story. Subsequent to Upakosala's recognition of his being full of desires and the fires' collective teaching, the three fires come forth individually and in turn instruct Upakosala. The three lessons have an identical structure: a list of four elements, a statement of self-identity (of each of the fires) and the fruits of meditation on this identity, including a vow of protection in this and the other world. The householder's fire, the *gārhapatya*, lists the following four elements: earth (*pṛthivī*), fire (*agni*), food (*anna*), and sun (*āditya*). It then makes the statement of identity: This person (*puruṣa*) who is seen in the sun, he is I, me, he is really myself' (*ya eṣa āditye puruṣo dṛśyate so 'ham asmi sa evāham asmīti*). This statement of the *gārhapatya* fire makes a characteristic Upanishadic demand on our (insufficient) inner space. It is a statement of identity between an objective entity (the person, *puruṣa*, seen in the sun) and the subjective 'me'. Upakosala seems to understand this statement sufficiently. He stays silent, and the *gārhapatya* fire continues with the fruits: 'Whoever knows it in this way and meditates on it, destroys evil deeds, becomes master of the world, lives in glory his full span of life, and his offspring never diminish.'[37]

Things seem to start looking up for Upakosala. His gloom dissipates, and he completes learning (understanding) the second lesson (the first one was the fires' collective teaching) from the *gārhapatya* fire, and moves on to the next teaching, from the *anvāhāryapacana* fire. This fire lists four entities: water, the quarters of the universe, the stars, and the moon, and moves on to the typical Upanishadic statement of identity: 'The person (*puruṣa*) seen in the moon is I, me, he indeed is myself' (*ya eṣa candramasi puruṣo dṛśyate so'ham asmi sa evāham asmīti*). And then the fruits available for one who knows and meditates: destruction of evil deeds, mastery of the world, full span of life, and undiminished progeny.

Upakosala gets deeper into the process of transformation by virtue of understanding what the fires tell him. Overcoming the crisis consequent

[37]ChU 4.11.2: *sa ya etam evaṃ vidvān upāste 'pahate pāpa-kṛtyāṃ lokī bhavati sarvam āyureti jyog-jīvati nāsyāvara-puruṣāḥ kṣīyante.*

to the recognition of desire, Upakosala makes room for the transforming teaching of the fires. His countenance brightens and an aura is visible around his head. Light and strong, he moves on. The third fire (*āhavanīya*) lists four entities: life-energy—*prāṇa*, space, heaven, and lightning, and repeats its own statement of identity with the person seen in the lightning.

But the fires are also conscious of their limitation and are thus somewhat apologetic; they cannot effect the complete instruction. We have taught you the knowledge about ourselves and the knowledge of the self. However, the teacher alone, they say, can show the 'way' (course) (*gati*). (*te hocur upakosalaiṣā somya te 'smad-vidyātma-vidyā cācāryas tu te gatiṃ vāktety*). Indeed, it seems that once again Upakosala experiences dismissal but this time by the fires, and probably undergoes a mini crisis of being rejected again. On the other hand, he makes room (or inner space) for more teaching. He must not be satisfied with the Vedic riches obtained thus far. Full of fear, he must now face the ultimate authority and undergo the final crisis as well.

The teacher returns home. It is a tense moment, according to Śaṅkara. The teacher might be very angry. The guru and his disciple exchange words of recognition: 'Upakosala,' says the teacher. 'Reverend Sir (*bhagavan*),' replies the disciple. The teacher perceives Upakosala's condition of enlightenment, and observes: 'Your head shines like that of a knower of brahman, my dear' (*brahmavida iva somya te mukhaṃ bhāti*). Then he enquires: 'Who instructed you' (*ko nu tvānuśaśāsa*)? At this moment, Upakosala gets deep into trouble. Will he confess the whole truth? His answer reveals anxiety, insecurity, but also deep attachment to his teacher. He introduces a defensive question: 'Who *could* teach me' (*ko nu mānuśiṣyād*)?[38] Indeed, the Upaniṣad reflects on Upakosala's question: 'He denies (suppresses)(*apanihnūte*) as it were (the teacher's apparent charge of betrayal).'[39]

Apparently unable to sustain the defensive mood of denying the truth, Upakosala continues to speak and, pointing at the fires, he utters a most enigmatic statement: *ime nūnam īdṛṣā anyādṛṣā iti*. 'They are actually such, like others.' Thus Upakosala seems to continue his denial of the whole truth, and yet, in his reference to the fires, he indirectly, through denial,

[38]M. Müller and P. Deussen understand the optative *anuśiṣyād as* 'should teach'. But if Upakosala denies that the fires had taught him, *anuśiṣyād* should be understood as could rather than should.

[39]Śaṅkara interprets Upakosala as saying: 'In your absence (*tvayī proṣite*), who could teach me?'

acknowledges the deeper truths of his anxiety, his hope of reconciliation, etc. In this interpretation, Upakosala communicates a tacit meaning opposite to that embodied in his overt speech. In other words, truth is manifested here more efficiently by untruth; untruth makes the teacher understand much and share his disciple's experience and relationship with his teacher. This sharing indicates a turning point in the narrative.

Some scholars see Upakosala's enigmatic statement as a clear-cut confession of the whole truth; thus Olivelle translates, similarly to Hume and Radhakrishnan: 'And alluding to the fires, he continued: "These look like this now, but they were different."'[40] P. Deussen and M. Müller leave the statement as it is: 'These (fires) are actually such, like others.'

Swami Swahananda emphasizes how fearful and intimidating the guru is, so much so as to deflect Upakosala's natural truthfulness. According to him, Upakosala means to say: 'The fires were shining before, but because of your coming, they seem afraid and trembling,' and saying this he points to the fires as his instructors. Upakosala also is afraid; so his behaviour should be construed as fear and not as concealment of the truth. That is why 'as it were' has been used.[41]

Mutual care and love are indeed hallmarks of the relationship of Upakosala with the fires. He takes care of them and they instruct him and provide him with the power to obtain Vedic goods such as long life, progeny, riches, fame, and security. And yet, in his progressive renunciation Upakosala says goodbye to the fires as well; Upakosala's ordeal, which is also his ultimate opportunity, nears its end. Satyakāma, the teacher, does not rebuke his student; neither does he leave him behind. Significantly, Satyakāma knows very well, from his own life experience, how difficult it is to say the truth on certain occasions.[42] Thus as Upakosala is seen trying to share false knowledge with his teacher through a deceptive question (what could the fires teach me?), his teacher seeks to share a moment of truth with Upakosala by assuming that both student and teacher know that the fires had taught Upakosala. The teacher apparently reads the truth through Upakosala's untruth and ignores the failing of his student,

[40] *Upaniṣads*, trans. by P. Olivelle, Oxford University Press, 1996, p. 134.

[41] The *Chāndogya Upaniṣad*, p. 307.

[42] See below the exposition of Satyakāma Jābāla's story according to the ChU 4. Thus, when he had applied for *brahmacaryatva* at his teacher's house, he was facing a truly devastating moment of choice. Asked about his lineage (*gotra*), he had to decide: will he cover up the awful truth about his mother and himself? Will he confess his ignorance about his origin (in particular his father's unknown identity)? Satyakāma met the challenge, and gained access to his guru's teaching.

overlooking the symptoms of the young man's weakness and deep crisis. He seems to realize how important he is to his disciple. Will Upakosala meet the new challenge? Will he respond to his teacher's grace? Upakosala's untruth is more valuable (and instructive) for Satyakāma. He provides Upakosala another opportunity and asks him: My dear, what did the fires teach you (*kiṃ nu somya kila te 'vocann iti*)? Upakosala admits: 'This' (*idam iti ha pratijajñe*). Apparently, the Upaniṣad interprets this moment as Upakosala's change of heart; the teacher moves on and defines the difference between the fires' teaching and his own. He says: 'While the fires taught you the worlds, I will teach you something else, much superior to this. As water does not stick to a stalk of lotus, so karma and evil do not contaminate (touch) one who knows my teaching (*ahaṃ tu te tad vakṣyāmi yathā puṣkara-palāśa āpo na śliṣyanta evam evaṃ vidi pāpaṃ karma na śliṣyata iti*).' In these words of the guru, the Upaniṣad distinguishes between the language of the fires and that of the teacher, the latter's being more heated and powerful. The fires' language is referential in a sense that the teacher's is not. The teacher's teaching bears fruits of a different order than those of the fires.

Now, Upakosala's speech-act admitting the truth creates or reveals his openness to his teacher's final instruction. It seems to open up a new zone of inner space. What does Upakosala *do* by his speech-act of *pratijñā*? What does Satyakāma do in his instruction? Upakosala's confession and Satyakāma's instruction are heated language events pointing at the centre of scorching understanding. There is a merger of teacher and disciple who share the moment of understanding.

Upakosala's harsh and unfathomable teacher is none other than Satyakāma Jābāla, who never knew his father. The *Chāndogya Upaniṣad* 4.4 tells his mother's and his story, a narrative well known by virtue of its bearing on the entitlement to study the Veda (*adhikāra*). Jabālā, Satyakāma's mother, was apparently a *śūdra* woman, a servant in her youth and most likely also at the time of the story. As a young boy, Satyakāma asks his mother's permission to leave home and live as a *brahmacārin* in a guru's house. He asks his mother about his lineage (*gotra*), and she answers: 'I don't know your lineage.'

Shrouded in painful uncertainty, Satyakāma opens up a painful—probably suppressed—area of experience in the family. Since in traditional Hindu society, the entire dharma-based structure of life (the *varṇāśrama* system) is grounded in parental identity, a family with unknown lineage is doomed to inferiority. Satyakāma triggers a big crisis, exposing the weightiest unfinished business possible in ancient India. Moreover, in his wish to leave his mother, Satyakāma challenges the person closest to him (maybe the only one) on her innermost being (her *śūdra*hood). It seems he is saying: 'Mother, I am not like you. You are a plain—even frivolous—servant, and I am different.' Thus, in his intention to leave home for a guru's house, Satyakāma dissociates himself from his mother. She is going to be left alone, rejected by her own son. The poor mother! Not only is her inner being as a *śūdra* rejected by her son, but her big secret as well: something is unusually wrong and unlawful in her past. In short, this is not going to be a nice, sweet, harmonious story.

The real narrative of the Satyakāma story—a narrative of crisis and bitter mother-son conflict—has never been read. Let us see how P. Olivelle sums up the story:

> In one of the most moving stories of the Upaniṣads, Satyakāma's mother, Jabālā, confesses to her son that he was born out of wedlock and that she does not know who his father is. She asks him to call himself the 'son of Jabālā' (Jābāla), thus adopting a matronymic. This open truthfulness so impresses his teacher, Haridrumata Gautama, that he initiates him into Vedic study (ChU 4.4–9).[43]

Olivelle—whose translation of the Upaniṣads is the most 'down to earth' to date—reduces the Satyakāma narrative to something so saccharine. In his translation, the mother confesses to her son that he was 'born out of wedlock' and she does not know who his father is. As if Satyakāma does not know that already! It is not likely that this is the first time that Satyakāma asks such a question; he is not an infant, for he is about to leave home and become a *brahmacārin*.[44] An obviously intelligent and curious boy, Satyakāma has certainly already noticed that he was born 'out of wedlock'. Satyakāma seeks a *real* confession from his mother, true information about his father, but he does not get it; for his mother tells him an untruth.

Olivelle's version—which is the conventional presentation of the story as a celebration of the great virtue of truthfulness—suggests a harmonious mother-son relationship, as Jabālā seems to trust her son so much that she tells him the whole awful truth about him and herself. And he, the son, takes in her story as a whole and quotes it word for word in his encounter

[43]*Upaniṣads*, trans. by P. Olivelle, Oxford University Press, 1996, p. 413.

[44]Satyakāma's age is not mentioned here. According to Manu, the age of disciples in a guru's house varies; a brahmin son is supposed to undergo *upanayana* at the age of eight, a *kṣatriya* at ten, a *vaiśya* at 11. How old then was Satyakāma? With no father (and identity) in sight, we do not know. If we knew Satyakāma's age at the time of his asking his mother, we could perhaps know about his perception as well as his 'real identity'.

with his prospective guru, Haridrumata Gautama. The latter is so impressed with the truthfulness shining through Satyakāma's quotation of his mother that he accepts him as a disciple; for who could stick to the truth but a brahmin? Olivelle's interpretation is essentially accepted by all commentators and scholars, yet a more contextual reading suggests that the real story which the *Chāndogya* tells is different.

Most likely, questions about his father surfaced much earlier in Satyakāma's mind. But the urge to ask his mother this time is apparently occasioned by his yearning for education and knowledge. Since his mother is apparently a *śūdra*, Satyakāma might have suspected that he is not entitled to knowledge of the Veda. And yet he asks. Why? He must have known he was a brahmin.

Interestingly, none of the translators and commentators says that Satyakāma's mother was a *śūdra*, according to the *Chāndogya*. No doubt, such a fact would change the reading of the Satyakāma story altogether. For if Satyakāma were suspected of *śūdra*hood, his application for Vedic studies in a guru's house would be subversive of the entire *varṇāśrama* system. See for example how Śaṅkara—in BSBh 1.3.38—lists the punishments inflicted on a *śūdra* caught hearing and studying the Veda:

> The *śūdras* are not qualified for that reason also that *smṛti* prohibits their hearing the Veda, their studying the Veda, and their understanding and performing Vedic matters. The prohibition of hearing the Veda is conveyed by the following passages: 'The ears of him who hears the Veda are to be filled with (molten) lead and lac,' and 'For a *śūdra* is (like) a cemetery, therefore (the Veda) is not to be read in the vicinity of a *śūdra*.' From this latter passage the prohibition of studying the Veda results at once; for how should he study the Scriptures in whose vicinity it is not even to be read? There is, moreover, an express prohibition (of the *śūdras* studying the Veda). 'His tongue is to be slit if he pronounces it; his body is to be cut through if he preserves it.'[45]

Śaṅkara's BSBh 1.3.38 refers also to the teacher's role. He must take care not to teach a *śūdra*. 'There are, however, express prohibitions also, such as "he is not to impart knowledge to the *śūdra*".'[46] Thus Satyakāma's encounter with his prospective teacher (Haridrumata Gautama) is charged with danger and risk on both sides. How dare Satyakāma, the *śūdra* woman's son, approach Haridrumata Gautama?

[45]See G. Thibaut (translator), *The Vedānta-Sūtras*, with the Commentary of Śaṅkarācārya, Oxford University Press, 1904, rep. by Motilal Banarsidass, 1994. See p. 77.

[46]Ibid.

But Śaṅkara, as we shall presently see, does not even touch upon the possibility of Satyakāma's mother being a *śūdra*. However, this is—most probably—what the Upaniṣad says. Let us see the way Jabālā presents herself to her son:

I do not know, my dear, what your lineage is. I was a maid in my youth, roaming about, and serving (others), and had many relationships; and (thus) I got you. My child, I do not know your lineage (*nāham etad veda tata yad gotrastvam asi. Bahvahaṃ carantī paricāriṇī yauvane tvām alabhe sāham etan na yad gotrastvam asi*).

Satyakāma, obviously, does not know who his father is; but he learns from his mother the circumstances of her youth (and his birth). The Upaniṣad does tell us that Jabālā was a *śūdra*. Satyakāma could thus suspect that he was dangerously close to being a *śūdra* himself, and thus not entitled to study the Veda. And yet he ventures to go to the guru's house, risking himself and the teacher. Why?

Before we continue with the real Satyakāma story, let us see more closely how the tradition—suppressing questions about Jabālā's possible *śūdra*hood, as we have seen—deals with the reality and meaning of Jabālā's youth. The most important source for consideration is, as usual, the commentary of Śaṅkara. He senses Satyakāma's apparent dissatisfaction over his mother's 'I don't know your lineage' answer, and in his presentation of the story Śaṅkara makes Satyakāma ask his mother: *How come* you do not know? (*kasmān na vetsi*), to which question the mother offers her version of her life story. But here of course it is Śaṅkara's presentation, his own—amazing indeed—view of Jabālā's whereabouts in her youth. According to Śaṅkara, she was married to a husband, in whose house (*bhartṛgṛhe*) she lived, and whose social identity she did not consider important enough to know (or remember)! 'My mind was not in thinking about things such as lineage' (*gotrādi smaraṇe mama mano nābhūt*). But whom did she serve so actively, roaming about, in her youth? The guests and visitors who entered her own family house. But—we must ask—who *was* the husband? She really does not remember, Śaṅkara suggests, and adds that at that time—when she was extremely busy with the guests—Satyakāma was born. In Śaṅkara's version, Jabālā tells Satyakāma: At that time your father died and I was desolate, this is how I was, and I don't know your lineage' (*tadaiva te pitoparataḥ ato 'nathāhaṃ, sāham, etan na veda yadgotraḥ tvam asi*). But who was Satyakāma's father?

Max Müller, who always heeds Śaṅkara's commentary, cannot accept Śaṅkara's description of Jabālā as her (mysterious) husband's wife, but he finds a way to save a little of her innocence. She was roaming about in her father's (not her husband's) house. He translates her speech thus: 'In my youth when I had to move about much as a servant (waiting on the guests in my father's house), I conceived thee. I do not know of what family thou art.'[47]

As seen above, Śaṅkara presents Jabālā as a respectable though strikingly obtuse woman, who does not pay attention to the most important fact of social life in India, namely, social identity. It is also remarkable that Śaṅkara says nothing of Jabālā's own identity; who is *her* father? What is her *gotra*? If not a *śūdra*, who is she? She could—at least—tell her son something about herself. And of course, even after Satyakāma's father dies, one can still go and find out who he was; he would surely have some relatives alive (or the guests and visitors could know). Thus, Śaṅkara seems to repress the obvious fact of Jabālā's identity; she is a *śūdra* woman, says the Upaniṣad. And yet, we still do not know who Satyakāma's father was.

It seems essential for Śaṅkara to delimit the scope of the challenge to the *varṇāśrama* system. As seen above, he is harshly clear-cut in his approval of the non-entitlement of *śūdras* to contact with the Veda. And yet, to some extent the Satyakāma story does attract attention (including Śaṅkara's) by virtue of its challenge to the *varṇāśrama* system. Even if Satyakāma is not a *śūdra*, he still does not know who his father is, and there is room for discussing the question of entitlement (*adhikāra*). Indeed, the issue of entitlement is the most important one for Śaṅkara and the entire subsequent tradition, in the context of the Satyakāma story. In his commentary on BS 1.3.37 Śaṅkara uses the term *liṅga*, an 'inferential sign' known from scripture (*śrutiliṅga*) to explain Haridrumata's decision to accept Satyakāma. W. Halbfass sums up the difference between Neo-Vedāntins and Śaṅkara on this matter:

> The representatives of the Neo-Vedānta usually consider this story of a young man who does not know who his father was and is classified by his teacher Haridrumata Gautama as a brahmin by virtue of his honesty to be an example of an ethical, characterological, non-hereditary view of the varna system. Śaṅkara, on the other hand, does not interpret Satyakāma's honesty as the cause and defining factor of his brahminness, but as a mere indicator of his hereditarily legitimate membership in the brahmin caste.

Thus the story of Satyakāma attracted the attention of those interested in the hereditary caste system and in the 'discrepancy between the

[47] *The Upaniṣads*, trans. by Max Müller, rep. Dover Publications, 1962, v.1., p. 60.

metaphysics of all-encompassing unity and the insistence upon strict hereditary barriers in the social domain and even in religious and soteriological matters.'[48] Indeed, if reality is one, and the metaphysical core of the individual is the Absolute (brahman), why bother about distinctions between people? Approached thus, the need to read the story vanishes in the light of the abstract and sublime metaphysics of the Upaniṣads. And so, as suggested above, the famous Satyakāma story from the ChU 4.4 *has been radically under-read.*

None of the commentators (including Śaṅkara) addresses the details of the story in order to decode the Upanishadic voice at this point. Most conspicuous is the avoidance of any reference to Satyakāma's mother's identity (as a *śūdra* woman). See for example how R.D. Ranade (1926) tells the Satyakāma story:

> In a famous passage of the Chāndogyopaniṣad we are told how Satyakāma, the son of one Jabālā, who had led a wanton life in her youth, asked his mother when he came of age, as to who it was from whom he was born, how the mother answered that she could only tell him that he was born of her though she was not quite sure from what father he was born, how when Satyakāma went to his spiritual teacher in order to get himself initiated, he was asked by the teacher as to what family it was from which he had come, how the youth Satyakāma gave a straightforward reply saying he did not really know from what family he had come, but that he only knew his mother's name, and that she had told him that she did not know from what father he was born, herself having led a very wanton life in her youth. 'Heigh!' exclaimed the spiritual teacher to Satyakāma, 'these words could not come from a man who was not born of a Brahmin. Come, I shall initiate you, because you have not swerved from the Truth.[49]

For Ranade, this Upanishadic story illustrates the value of 'Truth, the Supreme Virtue'. In Ranade's exposition there is no room for conflict between mother and son or teacher and disciple. Significantly, Ranade does not tell us how the teacher relegates Satyakāma to the wilderness for years to come, to care for four hundred feeble and lean cows; how Satyakāma seeks instruction elsewhere, and how he re-unites (like Upakosala) with his teacher, etc. Ranade shares Śaṅkara's avoidance of Jabālā's *śūdra*hood, though he pronounces her a 'wanton woman'. However, Ranade's account

[48]See W. Halbfass, 'Homo Hierarchicus,' in *Tradition and Reflection*, SUNY, 1990, p. 381.

[49]R.D. Ranade, A *Constructive Survey of Upanishadic Philosophy*, Oriental Book Agency, 1926, p. 311–12.

is on the verge of implying (or suggesting) that Jabālā was a *śūdra* (for he describes her as 'one Jabālā', an inferior and wanton woman). For him, the message of the story is the following: 'This story tells us how even the son of a wanton woman could be elevated to the position of a Brahmin merely for having told the pure and unadulterated Truth.'[50]

In this study, which focuses on the importance and meaning of the Upanishadic story, we must address the 'real story' here, namely, the secrets of Jabālā, Satyakāma's mother. More than 2000 years have gone by with Jabālā's secret intact. Allegedly bereft of bearing on the metaphysical and soteriological issues of the Upaniṣads, these secrets have not attracted attention. As seen above, the versions of the story are much at variance.

Does Satyakāma's mother tell him the whole truth? What is the voice of the *Chāndogya* at this point? There is an obvious note of evasion in her story. She was roaming about (*carantī*) in many places (*bahu*), serving, having many relationships, she says. But how many? Too many to know Satyakāma's father? Did she have intercourse with many men of many castes? Of so many castes as to not know which one (for in order to answer her son's question she only needed to know the caste)? A likelier possibility is that Jabālā is abandoned by a man of a higher caste, whom she does know as Satyakāma's father, and whom she does know as a man of a higher caste, most probably a brahmin, who does not marry her. We cannot know the heart of a *śūdra* woman involved with the life of a man of the highest caste in ancient India (eighth century BC). Her striking self-confidence in allowing her son to go to the guru's house is not as astonishing as it appears; she does know whose son Satyakāma is, and consequently she knows of her son's entitlement (*adhikāra*) to study the Veda in the preceptor's house. Satyakāma, in turn, knows his mother knows his father; this is why he asks the question in the first place. He comes to Haridrumata's house in the knowledge that he is a brahmin. In his turn, the guru collaborates with the secret, and declares Satyakāma a brahmin since 'he had not deviated from the Truth'. But what truth? Apparently, it is the horrible fact of not knowing his lineage. But the story is more complex than that; Satyakāma does know that his father was a brahmin, but he tells his teacher that his mother is a *śūdra* (and a 'wanton girl' in her youth). Śaṅkara is thus right to sense that Satyakāma was a brahmin by birth, and his truth-telling an 'inferential sign' (*liṅga*) of his brahminhood. But this is seen in the Upanishadic story itself; for the Upanishadic authority does not trust

[50] Ibid., p. 312.

the mother's account of her youth, and presents it as untruth told by mother to son, an untruth which confirms Satyakāma's inner conviction that he is by no means a *śūdra*.

Thus, we suspect Jabālā's account is a 'white lie', recognized as such by anyone who hears the story. But why suspect Jabālā's own account? Intercourse between man and woman, as a heightened, consequential contact, was of course limited and strictly regulated. However, *Manusmṛti*, the most important *Dharmaśāstra* text in ancient India, includes many references to the consequences of intercourse and the various offspring. It reflects upon kinds of marriage, kinds of inter-caste relations, etc. For example, Manu refers to twelve types of sons born to men and women. Let us browse a little through these possible types of sons, looking for the one most commensurate with Satyakāma's origin. 'The son that a man begets himself in his own field, in his legally married wife, he should recognize as his natural son, the first in rank.'[51] 'But the one whom a man makes his son, who is like him, knows the difference between right and wrong, and has the qualities of a son, he is to be recognized as a made son (*kṛtrima*)' (MS 9.169). 'And the one who is begotten in the house when no one knows by whom is the son secretly begotten in the house, and he belongs to the man in whose marriage-bed he was born' (MS 9.170). 'The one whom a man receives when he has been deserted by his mother and father or by either one of them is called the rejected (*apaviddha*) son' (MS 9.171). 'The son whom a priest begets out of lust (*kāmāt*) in a servant woman is a corpse (*śava*) who saves (*parāyan*), and so he is traditionally known as a "Saving-corpse" (*parāśava*)' (*yaṃ brāhmaṇastu śūdrāyāṃ kāmād utpādayet sūtam/sa parāyanneva śavastasmāt parāśavaḥ smṛtaḥ*)(MS 9.178).

The subtle relationship of truth and untruth: Jabālā's and Satyakāma's untruth—that she does not know the father—is decoded by Satyakāma, the guru and probably by Jabālā as well as the unspeakable truth of Satyakāma's brahminhood. But why does Jabālā hide the truth? Perhaps the man who is Satyakāma's father or members of his family are still around. We do not know. The unspeakable truth manifests itself through the untruth which is told. But what is the untruth? 'I do not know your lineage.' Jabālā's untruth thus implies the whole truth.

As works of art based on conflictual situations, endowed with irony and tension, the stories told in the Upaniṣads may not be as unsophisticated as they are invariably represented and interpreted. In Ricoeur's phrases

[51] *The Laws of Manu*, trans. by W. Doniger and B.K. Smith, Penguin Books, 1991, p. 216.

pertinent to the nature of symbols, the Upanishadic myths should be regarded as replete with the 'power of double meaning', or 'surplus of meaning', and as symbols they 'invite us to think' and 'never cease to speak to us'. Satyakāma's mother may not be as simple-minded and truthful as she is always represented to be. Moreover, Satyakāma's encounter with his teacher is—obviously—not as harmonious as it has invariably been represented. For indeed, the guru sends Satyakāma away. He re-enacts Satyakāma's essential ambiguity and sense of inferiority, sending Satyakāma away with four hundred thin (*kṛśa*) and weak (*abalā*) cows.

Significantly, Traditional Reading tends to ignore conflictual situations as well as 'inner splits' manifest in the lives of the Upanishadic heroes. Śaṅkara describes Satyakāma's site of exile as a forest (*araṇya*) full of grass and water (*tṛṇodakabahula*) and free of oppressing opposites (*dvandvarahita*). Such an interpretation painstakingly preserves reading into the story a sense of harmony between teacher and disciple, but at a cost: overlooking the rich texture of this narrative of conflict, inferiority, emergence of excellence and reaching knowledge. However, the Traditional Reading is not necessarily incoherent. One can see the possible meaning of Satyakāma's call to his teacher ('I won't be back without a thousand'); it may be an expression of sheer gratitude. But why then—a Good-Enough Reader would ask—does the narrator bother to describe the four hundred cows as lean and feeble?

The feeble cows are not healthy and whole. Similar to Satyakāma himself—found inexplicably lacking in eligibility to stay at the teacher's residence—the cows are apparently unworthy. The guru relegates Satyakāma and the cows to the wilderness (not necessarily a benign forest rich in grass and water). However, there is a dialogical quality to the scene of Satyakāma's expulsion. The guru says: 'Take care of these four hundred,' and Satyakāma responds: 'I will not return without a thousand.' The number four hundred is conspicuously insignificant, meaningless—as a number—in the Upanishadic literature. It is an 'inferior number', non-resonant, never in use. One thousand, however, is significantly different. We recall that King Janaka puts aside one thousand cows as prize for the sage who would go forth and prove himself best among the brahmins (*brahmiṣṭa*).[52] Jānaśruti brings to Raikva one thousand cows as part of the gift for gaining access to Raikva's knowledge and instruction.[53] Bālāki is

[52]BU 3.1.1.

[53]ChU 4.2. Significantly, Jānaśruti is rejected by Raikva during his first visit, when he brings along only 600 cows.

promised one thousand cows by Ajātaśatru in exchange for knowledge of Brahman.[54] Entitlement (*adhikāra*), access to superior knowledge (transformation) and excellence are associated with the number one thousand. Thus, in his offer of one thousand prospective cows, Satyakāma refuses to accept his guru's rejection and apparent judgement (or challenge). The transformation of the herd from four hundred into one thousand is Satyakāma's own transcendence of inferiority. And, of course, there may be a more earthly aspect—commensurate with the others—to the terse dialogue between teacher and disciple; in his present condition, Satyakāma may not be able to pay the teacher. His offer of future cows strongly suggests Satyakāma's recognition of his material inferiority.

We know that some years (*varṣagana*) have passed (a long time—says Śaṅkara); the Upaniṣad tells us. Satyakāma may not be so young now. As Satyakāma returns from the wilderness, in his journey to his master's house, he undergoes transformation. In his loneliness and apparent reluctance to reach for his teacher's instruction, he receives 'knowledge' from other sources. The first teaching is the bull's, preceded by the bull's suggestion that Satyakāma return home, since his pledge has been fulfilled.

Satyakāma's reunion with his teacher, after many years of separation and life in the wilderness, is the concluding climax of the story. Satyakāma returns whole, effulgent. Such a person, Śaṅkara says, is fulfilled (*kṛtārtha*), smiling, content, without worries (*niścinta*). In the course of his return to his teacher's house, a trace of ambiguity remains: 'You shine as if (*iva*) you are a knower of Brahman.' But then the teacher recognizes Satyakāma's enlightenment ('Who taught you?' he asks). And Satyakāma is fully in charge: 'non-human persons' (instructed me).

The correspondence of inferiority and instruction is most remarkable in the Satyakāma story. Being fulfilled and effulgent, inferiority-free, Satyakāma cannot receive new knowledge from his teacher. While the majority of the disciples portrayed in the Upanishadic stories receive knowledge which puts an end to their condition of inferiority, Satyakāma apparently receives from his teacher the same knowledge he has absorbed earlier from the bull, the fire, the goose and the water-bird. The fourfold teaching of Brahman transmitted by the four agencies signifies completion. One recalls in this context the famous fourfold nature of Brahman (*catuṣpad*) explicated in the *Māṇḍukya Upaniṣad* (a tetragonous explication which includes conscious merger with the Absolute—*turīya*). Under these

[54] KauU 4.1.

circumstances, the function of Upanishadic knowledge (transcendence of inferiority) can no longer materialize in the case of the whole, non-inferior disciple. Thus, the correspondence of type of inferiority to liberating Upanishadic knowledge is made explicit in the Satyakāma story. The initially disqualified, fatherless, initiated yet rejected student, the lonely leader of four hundred feeble cows in the wilderness, reaches the teacher's home (apparently his greatest moment) endowed and glowing with excellence and certainty. However, the connection of inferiority with knowledge is patently redundant here. Under these circumstances, the Upanishadic authority sees *no use* articulating anew a theory of the Absolute; such articulation would hardly count as an event in the story, bereft of the dialogical quality and value of contextual metaphysics. Indeed, Haridrumata's instruction to Satyakāma is unique and remarkable: it is not articulated at all. The Upaniṣad says: To him he said precisely this, having left nothing unsaid, having left nothing unsaid (*tasmai haitad evovācātra ha na kiñcana viyayeti viyayeti*). Śaṅkara emphasizes in this context that the teacher imparts to Satyakāma the same knowledge previously imparted to him by the 'deities' (*tam eva devatairuktaṃ vidyam*). In accordance with the inferiority/knowledge connection, the healing knowledge from the teacher's mouth—addressed to the non-inferior Satyakāma—is inherently, significantly devoid of innovation and force; it is indeed redundant, and left unsaid.

As its hermeneutic history shows, there is resistance to the real Satyakāma story. From a marginal perspective, Upanishadic stories have invariably been viewed as anecdotal episodes, or occasions for the real events of 'knowledge transmission', or as illustrations of higher matters (context-free metaphysics). As a concluding example of the under-reading supposition, let us turn to H. Oldenberg's presentation of the Upanishadic story. Oldenberg sees in these 'informative stories', consisting mainly of dialogues. This is his view of the use of the Upanishadic story:

> Men who converse here are obviously predominantly Brahmanas. The texts give a series of images of their life. They are short, clumsy sketches. But there is a waft of air in them which was breathed by the ancients. And we encounter, now venerable, now bizarre, phantoms of their own self, of Brahmanic thinkers, of those who hid themselves behind the secret-mongering verbosity, and also those, who having discovered the mystery of the Brahman, have found peace in it.[55]

[55]H. Oldenberg, The *Doctrines of the Upaniṣads and the Early Buddhism*, trans. by S.P. Shrotri, Motilal Banarsidass, 1991, p. 95.

Nice and peaceful indeed were the lives of the Upanishadic householders and seers of old, whom we 'imagine in the proximity of sacrificial places and grazing grounds with large cowherds, happiness and pride of their owners'.[56] And here is the Satyakāma story as Oldenberg reads it:

> A youth by the name of 'Satyakāma' reports to a teacher as his pupil. He asks him from what family he comes. My mother told me, she had moved about a lot in her youth; she does not know anything about my father. No one but a true Brāhmaṇa, says the teacher, is so honest. Fetch firewood; you may become my pupil.[57]

Thus, through a view from the margin, amidst grazing cows, loyal disciples and truthful mothers, we see a just and orderly universe, a world in which Satyakāma is recognized for what he is—a brahmin. And Upakosala as well, though somewhat tormented in the beginning (for he did not carry out all his duties) soon receives the ultimate teaching and his peace of mind is restored. As I have suggested in this chapter, the Upanishadic story may be read differently, and thereby seen to contain pronounced suggestions of conflict, crisis, and agony that are part and parcel of the teaching of the self and hence inseparable from any metaphysical content. Reading the under-read story and restoring the essential exegetical integrity of narrative form and content leads to a contextual metaphysics whose complexity, it seems to me, is the source of its redeeming potency.

[56] Ibid.
[57] Ibid.

THREE

Marginality and Great Moments: Contextual Metaphysics in the Story of Maitreyī

Upanishadic storytelling focuses on a variety of inferiorities and the subsequent paths to Brahman-knowledge and excellence. The inner space necessary for understanding thus consists of inferiority, structured pain. The stories represent a range of Upanishadic inferiorities, revealed in the Upanishadic narratives leading to the articulation of Brahman-knowledge and excellence. Śvetaketu's is the inferiority of pride, vanity inferiority. Upakosala's is the inferiority of desire (multiple desire). Satyakāma's is the inferiority of doubt (and entitlement, *adhikāra*). Nārada's is the inferiority of spurious knowledge. Naciketas' is the inferiority of mortality. Maitreyī's inferiority is also that of being mortal yet of a different kind, implied in the disintegration of the family, and her becoming a person. Such inferiorities have their universal flavour, yet they belong to Upanishadic times, circumstances, and meanings. The householder's infirmity, the disenchantment with desire, wealth, familial immortality (continuation through offspring), with the sacrificial fire and maintenance of universal connectedness, are all nourished by sensitivities grounded in ancient India of the eighth century BC. These sensitivities are composite; concomitant with the disintegration of old Vedic values, there emerges the attraction of Brahman, the subjective ground of reality and experience. The alluring call of this magnetic Absolute together with a certain distancing from old Vedic existence create Upanishadic inferiorities of many kinds. Such inferiorities and structured pains are not easy to connect with or understand. The Upanishadic narratives represent lively, powerful, hectic efforts at moving from inferiority to excellence through surrender to the call of a new possibility.

Glimpsing from a distance, margin-dwellers try to feel the heat of remedial intention, to experience the terrain on the other side of Upanishadic language. Dark and promising seem the whispers arising from secret communions in the Upanishadic crucible. Teachers, wives, disciples, gloomy householders, sages, fathers and sons (but not children) are being fermented in it, producing all the while utterances bound to become fateful for many. The Upaniṣads attest to great moments in the lives of those individuals who happen to be in the crucible, such as Śvetaketu, Maitreyī, Uddālaka, Yājñavalkya, Gārgī, Satyakāma, Upakosala, Nārada, etc. The texts which say something about these moments have become very important for the entire culture and identity of India, thus the Upanishadic great moments of heated communication and language must have been powerful indeed. Commentaries and scholarship have accumulated around these primordial events of therapeutic, transforming instruction.

According to these Upaniṣads, the great moments are particular and private moments. A father challenges his son; the latter is alarmed and anxious yet reconciled to his new sense of inferiority and seeks help; they participate in a discourse both intimate and prolonged (it goes on for days, weeks, or even months). Margin-dwellers, helplessly eavesdropping on the couple as they whisper, exclaim: Alas! This is nothing, after all, 'meaningless and a mere matter of words'.[1] A husband speaks confidentially to his wife (his other wife out of sight and earshot). Being childless, she is assertive—aggressive even—towards her inconsiderate, denying husband. Margin-dwellers are unable to feel her sorrow and anger, praising her incessantly for her humility and apparent spirituality, declaring her a 'fit consort' to her husband, who deserts her depriving her of offspring and 'immortality'.[2] A young boy revolts against his father, who sends him to his death.[3] A son challenges his mother for being a whore in her youth, asking her about his father.[4] Indeed, the great Upanishadic moments are often enveloped and developed within an intimate—and instructive—context. Indeed, the everyday intimacy of the Upanishadic story may reflect the nature of the Upaniṣads as esoteric, secret texts; A.B. Keith emphasizes that 'from an early period the utmost stress is laid on the secret character of the instruction contained in these works: the rule is that they are the highest mystery ...'.[5] It is a normal procedure for scholars and commentators to categorize the nature of the verbal exchanges reported in the Upaniṣads

[1]See Keith's judgement, pp. 62–4.
[2]See discussion of Maitreyī's story in BU 2.4 and 4.5, pp. 73–9.
[3]KU 1; see Ch. 4.
[4]See earlier discussion on Satyakāma's story (Ch. 2).
[5]A.B. Keith, *The Religion and Philosophy of the Vedas and Upanishads*, Harvard University Press, 1925, rep. Motilal Banarsidass, 1989, p. 489.

as secret and shrouded in mystery. Thus, for example, M. Hiriyanna observes: 'The word *upaniṣad* literally means "secret teaching" (*rahasya*) or the teaching which was jealously guarded from the unworthy.'[6] Upanishadic teaching is referred to as the 'secret teaching' (*guhya ādeśaḥ*),[7] 'ultimate mystery' (*paramaṃ guhyam*),[8] etc. And yet, the meaning and impact of this dimension of secrecy and mystery in the Upanishadic verbal exchange, seemingly close and accessible, almost familiar, is ignored by many scholars. M. Hiriyanna says:

> The classical Upaniṣads ... represent the flower of Vedic thought. They are written in rhythmic prose, where they are not metrical; and they possess, on the whole, a musical quality of their own. The reader, even when not familiar with their teaching, easily grasps their general import; and their power of transporting one out of oneself is remarkable.[9]

Hiriyanna's comment on the power of Upaniṣad-reading to transport the reader 'out of himself' seems to suggest that it is not so difficult. Music, indeed, often delivers such transport. Yet, in his assumption of easy access, Hiriyanna disregards the nature of Upanishadic communication, its secret, dialogical character. As a good margin-philosopher, his abstractions of the Upanishadic communications are lucid; he explains clearly the logical advantages of identifying *ātman* with brahman. And yet, he does not refer to any of the Upanishadic stories, nor does he mention any of the Upanishadic sages by name, or even refer to the 'meta-context' of the Upaniṣads in general (historical background, cultural milieu, etc.).

The secret nature of the Upanishadic exchange points to difficulties implied in the view from the margin to the centre. In addition to a formal definition of the Upanishadic teaching as 'mystery', the Upanishadic stories embody and reflect this feature. A sage invites someone to leave an assembly of scholars so that he (the sage) can say something important.[10] A sage confronts a colleague with the incredibly bitter and hateful charge: 'You are pretending to know, but you know not!'[11] Sages curse, humiliate and kill one another.[12] A disciple confesses forbidden love to his master's

[6]M. Hiriyanna, *The Essentials of Indian Philosophy*, George Allen & Unwin, 1985, p. 18.

[7]See ChU 3.52.

[8]See KU 1.3.17.

[9]M. Hiriyanna, *The Essentials of Indian Philosophy*, George Allen & Unwin, 1985, p. 18.

[10]See BU 3.2. The discourse unfolds between Yājñavalkya and Arthabhāga.

[11]See BU 3.7. Uddālaka makes the charge against Yājñavalkya: 'Everyone can say I know, I know. Say if you know!'

[12]See Uddālaka's threat to Yājñavalkya (BU 3.7), Yājñavalkya's curse to Śākalya

wife;[13] a mother lies to her son about his father.[14] These are all intimate, charged moments which develop into occasions for expressing redeeming, therapeutic truths. Traditional truth-seeking scholarship mistakenly discards these stories as shells as they seek kernels which they then identify as teachings. This 'peeling-to-a-core' reading strategy then determines (and limits) interpretation.

The great Upanishadic moments are often featured as awakenings within specific relationships. Are these relationships ('context') really important? I believe they are. Thus for example, if Uddālaka's teaching to his son is related to the latter's pride and the father's compassion, the teaching may remedy pride, though not at the expense of one's essential individuality. The *Tat tvam asi* message, taken as contextual metaphysics, might change one's ego boundary to some (unknown as yet) extent, without necessarily depriving one entirely of 'individuality'. Thus the context within which the 'message' finds expression may determine the scope and nature of transformation implied by the text, and scholarly inattention to context, to the situatedness of the eagerly sought 'message', invites distortion like that implied in the ease with which Hiriyanna extracts 'their general import'. Ignoring narrative demands in favour of textual excavation, these scholars recontextualize the reception and interpretation of the intimate whispers arising from the transforming crucible.

The choice of context is a significant dimension of the margin. Philosophers, narcissistic seekers, historians, missionaries, peer from the margin at the Upanishadic great moments. What do these margin-dwellers see? To classify different locations on the margin is not an easy task; many are the onlookers of the great moments. Though spectators from the margin, they are not casual intruders. Rather, they share a sense of the value and importance attached to the moments they desperately watch and seek to understand, praising them from different perspectives and locations on the margin. Gloomy Schopenhauer is a particularly outspoken, enthusiastic margin-dweller with respect to the value and role of the Upaniṣads in his life. His follower, the great Indologist P. Deussen, chooses to quote Schopenhauer's statement as the motto of his (Deussen's) translation of *Sixty Upaniṣads of the Veda*: [15]

(BU 3.9). Also J. Brereton, 'Yājñavalkya's curse', and my article 'The Upanishadic Story and the Hidden *Vidyā*', *Journal of Indian Philosophy*, 1998, pp. 373–85.

[13]See Upakosala's story (ChU 4.10) discussed on pp. 38–41.

[14]See the story of Satyakāma Jābāla and his mother, ChU 4.4.

[15]See W. Halbfass, *India and Europe*, SUNY, 1988, p. 106. Halbfass' discussion on Schopenhauer and the Upaniṣads is penetrating and exhaustive.

How every line is full of sure, definite and throughout harmonizing significance! How out of every page confront us deep, original, elevated thoughts, while a higher and highly sacred earnestness vibrates through the whole! Everything here breathes forth the Indian atmosphere and primordial existence akin to nature! And, O! how here the mind is washed of all Jewish superstitions which it had formerly cherished! It is the most rewarding and the most elevating book which (excepting the original text) there can possibly be in this world! It has become the solace of my life and will be the solace of my death.[16]

Schopenhauer's exclamations are vibrant and sincere. He finds in the Upaniṣads a true echo of his own philosophy and ultimate wisdom, since 'in general, the sages of all times have always said the same',[17] and 'Buddha, Eckhart, and I all teach essentially the same'.[18] However, Schopenhauer 'never made an effort to learn Sanskrit, although he repeatedly glorified the excellence of this language and the rewards of mastering it'.[19] Although he praises the Upanishadic authors, who 'could hardly be thought of as mere mortals', he does not mention any (such as Yājñavalkya and Uddālaka Āruṇi) by name. He is conspicuously cautious, and restricts himself to the reading of Anquetil Dupperon's Latin translation of fifty Upaniṣads (*Oupnek'hat*, 1801/2) from the Persian version of 1656 (under Dara Shukoh). Other translations were available in Schopenhauer's lifetime, but he does not seem to have chosen to look at them. He explicitly saw in the Upaniṣads 'his own teachings reflected in them "as in a mirror"'.[20]

Schopenhauer is an example of a particularly idiosyncratic location on the margin. The affinities between his teachings and some of the ancient Indian doctrines often seem striking indeed, and his unreserved acceptance of the Upaniṣads—with no compunction about hermeneutic complexities in the East-West encounter—can seem 'correct', almost justified.

Other commentators sound poignantly aware of the difficulties involved in attending to the great moments while dwelling on the margin. C.G. Jung notes that '[a] beggar is not helped by having alms, great or small, pressed into his hand, even though this may be what he wants. He is far better helped if we show him how he can permanently rid himself of his beggary by work. Unfortunately, the spiritual beggars of our time are too inclined to accept the alms of the East in bulk and to imitate its ways unthinkingly.'[21]

[16] P. Deussen, *Sixty Upaniṣads of the Veda*, trans. by V.M. Bedekar and G.B. Palsule, Motilal Banarsidass, 1991, p. vi.

[17] Cf. W. Halbfass, *India and Europe*, SUNY, 1988, p. 111.

[18] Ibid.

[19] Ibid., p. 106.

[20] Ibid., p. 111.

[21] C.G. Jung, *Jung of the East*, edited by J.J. Clarke, Routledge, 1995, p. 45.

To his fellow dwellers-on-the-margin, Jung advises: 'What it has taken China thousands of years to build cannot be acquired by theft. If we want to possess it, we must earn the right to it by working on ourselves.'[22] Working on himself, it seems, produced Jung's identification of Brahman with the libido.

> Even man's strength comes from *brahman*. It is clear from these examples, which could be multiplied indefinitely, that the *brahman* concept, by virtue of all its attributes and symbols, coincides with that of a dynamic or creative principle which I have termed libido.[23]

While Schopenhauer and Jung offer us paradigmatic glimpses of the problems involved in peeping at Upanishadic language, its otherness and heat, it is obvious that they are not scholars. However, it seems that the most populous location on the margin is not that of the self-enveloped psychologist, the spiritual seeker or the gloomy narcissist; the philosopher and philologist, the scholarly explorer's has become the most haunted site of people peeping at the Upaniṣads.

Of this congregation, A.B. Keith is not only a great scholar, but also a courageous thinker. While many margin-dwellers—in particular scholars like himself—refrain from addressing the issue of truth-value or validity of the Upanishadic metaphysics—an issue awfully frightening for ordinary margin-dwellers—Keith is bold and admirably straightforward in the questions he raises:

> What judgement are we to pass on the main thought of the Upaniṣads? Are they really the expressions at an early date of the deepest principles of philosophy, or are they merely of historical value, interesting pictures of the early thought of man, important not *per se*, but because of their dominating influence on the future philosophy and life of India?[24]

Keith moves on, addressing two available answers (or better, orientations) with respect to the issue of the truth and value of the Upanishadic teaching. Keith presents A.E. Gough as a scholar who is essentially a historian denying the Upaniṣads' ultimate value and truth. The spiritual and intellectual inferiority of the Upaniṣads is explained by the 'admixture of blood among the Indians, and the corrupting influences of a low

[22] Ibid.

[23] Ibid., p. 145.

[24] A.B. Keith, *The Religion and Philosophy of the Veda and Upanishads*, Harvard University Press, 1925, rep. Motilal Banarsidass, 1989, p. 592.

order of civilization'.[25] But Keith is not easily diverted from his search for the truth and value of the Upaniṣads. In his truly philosophical mood he invokes another orientation, that of P. Deussen who—as Keith sums it up—says that 'Yājñavalkya had anticipated the view of Schopenhauer regarding immortality as indestructibility without continued existence.' Keith summarizes Deussen's motivation and purpose:

> His view of the Upaniṣads is, therefore, greatly determined by his double desire to find in them the anticipation of Kant and to show that the views of Kant, as modified by Schopenhauer, are the only possible views of philosophy.

Thus, Keith diagnoses—with tacit respect and admiration—the condition of his close fellow dweller-on-the-margin.[26] Similarly to Deussen, he partially accepts the latter's criterion of truth (compatibility with Kant's metaphysics), but he also goes beyond it. Bold and committed to philosophy, independently of Deussen, Keith continues in his own way to address the ultimate question: what is the truth and what is the value of the philosophy of the Upaniṣads? Along with Deussen, Keith distinguishes between two components in the Upaniṣads; the abstract (philosophical) part and the theories taken from the Brāhmaṇas. The latter are, in Keith's view, 'of historic interest: as philosophy, they are unworthy of a moment's consideration'.[27] Keith seems to overcome some reservations he also has with respect to the more abstract and philosophical portions of the Upaniṣads ('the ideas of Yājñavalkya'). He finally seems ready to approach them: 'It is, therefore, necessary to consider these doctrines in themselves in their value for thought.'[28] After discussing the nature of the self—in principle unknowable—in the philosophy of the Upaniṣads, he gives his reasoned verdict:

> But the identity of the self and the absolute is based merely on the abstraction of the self as subjectivity, and that of the absolute as subjectivity, and the identity is therefore meaningless and a mere matter of words.[29]

[25]Ibid.

[26]Ibid., Keith finds much value in Deussen's motivation and presuppositions, and addresses the issue which he sees as central to Deussen's orientation. Thus, he says: 'The more important question which arises is to what extent the doctrines of the Upaniṣads do foreshadow the views of Kant' (p. 592).

[27]Ibid., p. 593.

[28]Ibid.

[29]Ibid.

Thus, Keith says that the Upanishadic teaching of the self and the absolute is essentially a tautology, 'a mere matter of words', as he says. 'Subjectivity is subjectivity,'[30] he suggests, is the Upanishadic message. And indeed, something rings true in Keith's argument; the Upanishadic whispers are reflected and refracted on the margin—the logician philosopher's, Keith's location on the margin—as tautologies. Similarly to Hiriyanna's exposition, Keith's is very lucid, and we do feel that we understand what he says. And yet, there is also room for suspicion ('thinking'). Is not Keith's aggressive abstraction of the Upanishadic teaching ('subjectivity is subjectivity') somewhat too aggressive? Can the Upanishadic tautology and 'mere words' transport people out of themselves? Whence the strong sense of incomprehension associated with reading the Upaniṣads? I suggest that Keith has left something out, and that this something is the Upanishadic story he so readily discards. His cursory reading of the Upanishadic stories (discussed later) then determines his (context-free) interpretation of the Upanishadic philosophy. Elucidating the stories while favouring lucidity over ambiguity, Keith's reading is already partial at its inception.

In a sense, Keith's judgement is the essence of marginality, the end result of a process of translation of the centre to the margin whereby ambiguity is relentlessly clarified. Even if thinkers on the margin (such as Schopenhauer) are often more enthusiastic about the meaning and value of the Upanishadic thought and religious experience than Keith's 'mere words' suggests, they are bound to marginal lucidity in essential ways. Perhaps the most important one is the level and mode of articulation ascribed to the hot language emanating from the abyss. In the case of the lucid philosopher with a low threshold of ambiguity—Keith is a good example—the expectation, processing, and absorption of the Upanishadic language is ineluctably reductive.

Endowed with his own inner space—more complex than just a logician's—Keith approaches the Upanishadic teaching from philosophical as well as from other angles. He doubts the efficacy of Upanishadic knowledge in the lives of people: 'It is simply inconceivable why on the ground of such theoretical knowledge men should abandon the desire for children, should give up their property, and wander about like beggars, practising a foolish asceticism.'[31] Having passed his judgement

[30] This summary of Keith's notion of the Upanishadic equation of *ātman* and brahman is mine.

[31] A.B. Keith, *The Religion and Philosophy of The Veda and the Upanishads*, Harvard University Press, 1925, rep. Motilal Banarsidass, 1989, p. 595.

on the core of the philosophy of the Upaniṣads, he also complains about the theory of maya which 'turns the supreme lord into a conjurer',[32] and the 'doctrine of Karman' wherein 'the hopeless inconsistencies of the view of Yājñavalkya become painfully obvious'.[33]

Unlike Keith, the great scholar C. Lanman does not care much about Upanishadic truth; he rather seems alarmed by the possible impact of the Upanishadic perspective upon the humane and soft values infused on Western civilization by the 'gentle Nazarene':

What a prospect, dark and void—this Supreme Spirit, before whom all human endeavour, all noble ambition, all hope, all love, is blighted! What a contrast, a relief, when we turn from this to the teachings of the gentle Nazarene![34]

Concern for the consequences of this kind of cross-cultural contact is not limited to Lanman, just as reductive engagement is not exclusive to Keith. Lives on the margin present us with many kinds of thirst, talent, disposition, education, and experience (biography). Many indeed are the seekers, philosophers, missionaries, travellers, historians, Neo-Hindus, and poets peering from our populated margin.[35] All are stooping over the dark abyss of the 'great moments' from which emerges language more or less hot.

Listening with a higher threshold of ambiguity to the dark abyss, we can recognize 'great moments' rather than 'great sayings' and great texts. Language in its different powers is indeed an important element of the abysmal great moments, but similar to the great moments themselves, it is tangible, material. Part of the tangibility and power of this language is recoverable in the Upanishadic story, heretofore dismissed as mere truth-container with the single function of conveyance of Upanishadic truth, discarded upon arrival.

The Upanishadic stories are more or less known. Margin-dwellers know of the discourse of Uddālaka Āruṇi with his son (ChU 6), of Indra's quest for the knowledge of the *ātman*, understanding which satisfies all desires (ChU 8.7), of Yājñavalkya's competition with his colleague sages

[32]Ibid.

[33]Ibid.

[34]Ibid., Keith quotes from C. Lanman, *The Beginnings of Hindu Pantheism*, Charles W. Sever, 1890, p. 24.

[35]W. Halbfass' *India and Europe*, SUNY, 1988, is a deep and exhaustive survey and analysis of the variety of encounters between India and Europe. Halbfass, however, does not use the metaphor of the margin and the centre.

(BU 3), the story of Satyakāma Jābāla's acceptance by his teacher (ChU 4.4) and the story of Maitreyī and her husband, Yājñavalkya (BU 2.4 and 4.5). Other stories such as that of the disciple Upakosala, tormented by desire (ChU 4.10), and the tale of Uṣasti Cākrāyaṇa and his 'child-woman' starving in the village of the 'elephant-owner' (ChU 1.10), are probably less known. However well known an Upanishadic story, its potential for recovering some of the 'great moments' and its relevance for the 'contextual metaphysics' of the Upaniṣads is invariably overlooked.

Maitreyī's story occurs twice in the oldest Upaniṣad, the *Bṛhadāraṇyaka Upaniṣad* 2.4.1 and 4.5. Indeed, this is a particularly well-known Upanishadic story. It is justified to note that Maitreyī's 'conversation with her husband, repeated twice, is one of the more important sections of the BU',[36] or even agree with Oldenberg, that 'the conversation between Yājñavalkya and his wife Maitreyī can be called the culminating point of the Upaniṣads'.[37] There are quite a few differences between the two versions of the story. P. Deussen suggests that the two versions come from two different schools, united at some point. (Deussen prefers the BU 2.4 version.)[38]

Yājñavalkya has two wives, Kātyāyanī and Maitreyī. Maitreyī is interested in discourses about the absolute (*maitreyī brahmavādini babhūva*), while Kātyāyanī is more knowledgable about women-business (*stri-prajñaiva tarhi kātyāyanī*). Yājñavalkya now readies to move away into a different mode of life (*atha yājñavalkyo 'nyad vṛttam upakariṣyan*). Being married, Yājñavalkya is described in this context as a householder who is about to leave for the wilderness. He calls on Maitreyī and says: 'I am about to wander away from this condition (*sthāna*). Let me make a settlement (*ānta*) (of property) between you and that one, Kātyāyanī' (*pravrajisyan va are 'ham asmāt sthānād asmi. hanta, te 'nayā kātyāyanyā 'ntaṃ karavāṇi*). Maitreyī says: 'Suppose I possess this whole world filled with riches, would it make me immortal?' (*yan nu ma iyaṃ bhagoḥ sarvā pṛthivī vittena pūrṇā syāt, kathaṃ tenāmṛtā syām*). Yājñavalkya answers: 'No, no' (*neti neti*).

Maitreyī says: 'What is the use of something which cannot make me immortal? Tell me just what you know' (*yenāhaṃ nāmṛtā syām kim ahaṃ*

[36] P. Olivelle (translator), *Upaniṣads*, Oxford University Press, 1996, p. 410.

[37] H. Oldenberg, *The Doctrine of the Upaniṣads and the Early Buddhism*, trans. by S.B. Shrotri, Motilal Banarsidass, 1991, p. 56.

[38] See P. Deussen, *Sixty Upaniṣads of the Veda*, trans. by V.M. Bedekar and G.B. Palsule, Motilal Banarsidass, 1991, p. 501.

tena kuryām̐ yad eva bhagavān veda tad eva me brūhīti). And Yājñavalkya says: 'You are really dear to me, and you speak dear words, and now my fondness of you has even increased! I will tell you this. But you heed well (concentrate on) my instruction' (*etad vyākhyasyāmi te vyācakṣāṇasya tu me nididhyāsasveti*). And Yājñavalkya goes forth to teach his wife the lesson she seemingly asks for: 'The husband, my lady, is not dear for the mere desire for husband, but the husband is dear for the desire for oneself (or one's self)' (*na vā are patyuḥ kāmāya patiḥ priyo bhavaty ātmanas tu kāmāya patiḥ priyo bhavati*). This is obviously an enigmatic statement; this enigma is, however, repeated in many similar statements (followed by many more illustrations). In the course of his instruction, Yājñavalkya teaches that man, once dead, has no consciousness (or awareness, *saṃjñā*) anymore (*na pretya saṃjñāsti*). 'That's what I say, my lady,' he emphasizes at the end of this long session. But Maitreyī, silent throughout her husband's talk, is confused, even frustrated. She says: 'You have confused me at this point, by saying that when man dies there is no more consciousness' (*atraiva mā bhagavān amūmuhāt, na pretya saṃjñāstīti*). And Yājñavalkya responds: 'My lady, I have not said anything confusing' (*na vā are 'haṃ mohaṃ bravīmi*), adding that this (what he has already said) is 'enough for understanding' (*alaṃ vā ara idaṃ vijñānāya*).

Scholars invariably see in this story the epitome of harmony and love between husband and wife. 'I love you even more by now,' says Yājñavalkya; but should we believe him? Should we take the Upanishadic voice (narrator and narrative) here at 'face value?' What indeed is the story? What is the—crucially important—subtext?

All commentators and scholars focus on the harmony and love prevailing between Yājñavalkya and Maitreyī; 'The sage Yājñavalkya, in a celebrated dialogue with his beloved wife Maitreyī, states that for the released and perfect knower there is no consciousness following death ...'.[39] While Yājñavalkya's speeches exhaust all scholarly and commentatorial attention, what is left of the story is the usual emphasis on Maitreyī's spiritual nature and her openness to her celebrated husband. Thus Radhakrishnan tells Maitreyī's story:

> Spiritual inclination is essential for the pursuit of spiritual life. In the *Bṛhadāraṇyakopaniṣad*, Yājñavalkya offers to divide all his earthly possessions between his two wives, Kātyāyanī and Maitreyī. The latter asks whether the whole world filled with wealth can give her life eternal. Yājñavalkya says: 'No, your life

[39]H. Zimmer, *Philosophies of India*, Meridian Books, 1957, p. 362.

will be just like that of people who have plenty of things, but there is no hope of life eternal through wealth.' Maitreyī spurns the riches of the world remarking, 'What shall I do with that which will not make me immortal?' Yājñavalkya recognizes the spiritual fitness of his wife and teaches her the highest wisdom.[40]

Now, if Maitreyī is just an open-minded, spiritually gifted woman who seizes an opportunity to receive instruction, the connection of the story with the contents of instruction is superficial, flimsy, unnecessary. If Maitreyī is not distressed or in pain, her application for knowledge is merely scholarly luxury. In this case, the only trigger and narrative justification for the unfolding of Yājñavalkya's theory of self would be Maitreyī's entitlement to the teaching, an entitlement deriving from her spirituality (expressed in her choice of knowledge over property):

There was no need for this heroic soul (Maitreyī) to give further proof of her spiritual quality. As always, Yājñavalkya rises to the occasion, and the tone of his discourse on Love and the Soul, on death and immortality, is on a plane with his other great discourses.[41]

But in this approach, the circumstances in which the teaching is embedded—as specified in the story—are too weak to shed any light on the teaching and its meaning. The story is thus not only aesthetically inferior but inefficient as well; after all, the nature of the classroom is only of meagre assistance in understanding the bright philosopher's lecture.

But we need not assume that this is the case in the Upanishadic story and text. In the case of BU 2.4 and 4.5, the teaching is ambiguous and enigmatic, and it is embedded in a story of deathly distress, pain, and crisis. Maitreyī learns of her weakness, a kind of inferiority. As the family disintegrates, she becomes a person. Becoming a person is thus Maitreyī's story. Under the new circumstances of her husband's leaving, emerges her plea for 'immortality' (*amṛtatva*). Her inferiority is thus explicitly defined in the Upaniṣad as that of being mortal, a person. It is, the story tells us, a hole unknown, an open space of pain to be relieved by 'knowledge'. 'Tell me what you know,' she says. She, indeed, is the real heroine of the story, facing a crisis—as Jānaśruti, Satyakāma, Nārada and others do—and seeking remedial knowledge.

Thus, although the story of Maitreyī sounds sweet to many, as husband and wife seem to be in perfect harmony, there are evidently some areas

[40]*The Principal Upaniṣads*, edited by S. Radhakrishnan, George Allen & Unwin Ltd., 1953, p. 101.

[41]P.D. Mehta, *Early Indian Religious Thought*, Luzac & Company, 1956, p. 127.

of conflict between these two spiritual persons. Moreover, as we watch Yājñavalkya and Maitreyī more closely, it becomes evident that conflict and crisis underlie Maitreyī's story to such an extent that, rather than a story of happy communion and spiritual elevation, Maitreyī's story is a tale of terrible conflict, anger, misunderstanding, and—for Maitreyī, at least—an unhappy ending.

Yājñavalkya leaves for the wilderness. The Upaniṣad does not tell us why. Moreover, the Upaniṣad remains silent about the most important component of the narrative; Yājñavalkya leaves his wives behind without a son. This condition of Yājñavalkya and Maitreyī being without offspring at the point of the former's departure is the vital subtext of the story. It is inconceivable, we might say, that in Upanishadic India such a condition—no offspring—could pass unnoticed or be disregarded.

According to Manu, a householder is entitled to leave for the wilderness under the following conditions, specified in the beginning of the sixth chapter of *Manusmṛti*:

6.1: After he has lived in the householder's stage of life in accordance with the rules in this way, a twice-born Vedic graduate should live in the forest, properly restrained, and with his sensory powers conquered.

6.2: But when a householder sees that he is wrinkled and grey, and (when he sees) the children of his children, then he should take himself to the wilderness.

6.3: Renouncing all food cultivated in the village and all possessions, he should hand his wife over to his sons and go to the forest—or take her along.

6.4: Taking with him his sacrificial fire and the fire-implements for the domestic sacrifice, he should go out from the village to the wilderness and live (there) with his sensory powers restrained.[42]

Moreover, Manu explicitly prohibits the quest—in the wilderness—for liberation (*mokṣa*) before one has paid his three debts (to the gods, teachers, and ancestors):

6.35: When a man has paid his three debts, he may set his mind-and-heart on Freedom; but if he seeks Freedom when he has not paid the debts, he sinks down.

6.36: When a man has studied the Veda in accordance with the rules, and begotten sons in accordance with his duty, and sacrificed with sacrifices according to his ability, he may set his mind-and-heart on Freedom.

[42]*The Laws of Manu*, translated by W. Doniger with B.K. Smith, Penguin Books, 1991, p. 117.

6.37: But if a twice-born man seeks Freedom when he has not studied the Vedas, and has not begotten progeny, and has not sacrificed with sacrifices, he sinks down.[43]

Manu's narrative of the householder leaving for the forest is a complex one (discussed later), and it reflects—as P. Olivelle explicates—a complex solution to a basic conflict of the Indian tradition.[44] For our purpose here, the reading of the (much under-read) story of Maitreyī, it is sufficient to say: Manu's laws concerning the transition from the *gārhasthya* stage of life (*āśrama*) to the next (wilderness-dweller, *vānaprastha*) are paradigmatic in Brahmanical India, and shed light on the basic situation suggested in the BU 2.4 and 4.5; a major conflict of tradition manifests in the life of a married couple. Yājñavalkya leaves Maitreyī (and apparently his other wife as well) without having begotten children (sons) with her, thus opening a major conflict with the tradition and values of Vedic society. Indeed, in the entire relevant literature (Brāhmaṇas and Upaniṣads), we see no reference to Yājñavalkya's offspring. His life-story is thus conspicuously different from that of his great colleague Uddālaka (who has Śvetaketu). Thus the most important story of the *Bṛhadāraṇyaka Upaniṣad* reflects a situation of cardinal crisis, as Yājñavalkya leaves for the wilderness before the fulfilment of one of the central duties prescribed for Vedic man: begetting a son.

As we shall presently see, the notion of immortality underlies Yājñavalkya's and his wife's shared distress; an offspring in Vedic and Brahmanical India was closely associated with immortality. Numerous indeed are the references to this association (of offspring and immortality), such as that in *Ṛg-Veda* 5.4.10, which expresses the prayer: 'Through offspring, O Agni, may we attain immortality.'[45] Thus says P. Olivelle:

The Brahmanical conceptions of immortality as freedom from physical death and of the family as the true and complete person are reflected in the belief that a man's immortality is found in his son. The family line continues in the son despite the death of the father. As the son survives after the father's death, so the father in his son survives his own death ... the family is what guarantees human immortality.[46]

Now, what is the meaning of this absence of a son in the story? Indeed, the story itself tells us. Yājñavalkya says that he is going to settle matters

[43]Ibid., pp. 120–1.

[44]*Sannyāsa Upaniṣads*, trans. with an introduction by P. Olivelle, Oxford University Press, 1992, pp. 19–81.

[45]Ibid., p. 27.

[46]Ibid.

between Maitreyī and Kātyāyanī, and Maitreyī asks: 'Will it make me immortal?' All interpretations hitherto focus on Maitreyī's desire for ultimate knowledge—not connected with absence of offspring—as the motive behind her quest for immortality. But the Upanishadic story offers us a different perspective. Maitreyī addresses neither the prospective division of property or knowledge. She is interested neither in the property nor in knowledge; rather she says: 'You are leaving for the forest without giving me a son!' Thus, Maitreyī actually asks: 'Without a son, how can property make me immortal?' And Yājñavalkya must concede: 'Property cannot make you immortal; it can only make you rich.'

Is the crisis resolved? Obviously not. For Yājñavalkya is very badly at fault. A most painful crisis unfolds here; a childless woman, a sonless man and a marital crisis. This crisis is perhaps an expression of the tension between celibacy and marriage in ancient India (the major 'conflict in tradition').

But the childless Maitreyī is not the only one in distress, left alone—prematurely—by her husband; Yājñavalkya himself renounces his mode of Vedic completeness. 'Next to sacrifice, therefore, the obligation to get married and procreate children (especially a son) is central to Brahmanical theology, which regards the family—father, mother, and son—as the only complete person.'[47] Thus, the sonless Yājñavalkya forsakes a major opportunity for wholeness and immortality in leaving for the wilderness. Moreover, the composition of the BU suggests that the 'theology of debt' must have been acutely present in Yājñavalkya's mind: 'In the opinion of our authors one must, above all, be quit of the sacred triple debt: towards the sages by studying the Vedas, towards the gods by performing sacrifices or religious rites, and towards one's ancestors by founding a family and having children to continue the cult of the *pitṛs* (forefathers).'[48]

Thus, the crisis of Maitreyī and Yājñavalkya is complex; both are deeply troubled and in need. While Yājñavalkya is somewhat better off—since he perhaps envisages the prospect of therapeutic knowledge and a different type of immortality—Maitreyī must be most bitter and helpless. And yet, the undeniable expressions of care on Yājñavalkya's part may reflect the shared anguish of this couple.

Maitreyī, as seen above, demands that Yājñavalkya give her his knowledge. This is perhaps a bitter, angry supplication. She demands indeed what had been an impossibility, since Yājñavalkya has never instructed

[47]Ibid., p. 26.

[48]R. Lingat, *The Classical Law of India*, University of California Press, 1973, p. 50.

his wife before. And since the Upaniṣad emphasizes the fact that Maitreyī is an intelligent woman, interested in 'theological discourse' (*brahmavādinī*), it is likely—according to the Upanishadic story—that she has wished previously to share her husband's knowledge. Up to the present moment of Yājñavalkya's departure for the forest, he has consistently rejected her supplications. He probably considered his wife not entitled to receive instruction about the *ātman*. It is Maitreyī's (and his own) location in crisis which apparently moves Yājñavalkya to make the attempt to instruct his wife. The tokens of Yājñavalkya's ambivalence towards teaching his wife are obvious. Kātyāyanī's role in the story also contributes to our understanding of Yājñavalkya's rebuke to Maitreyī. Though Maitreyī is *brahmavādinī*, she is also a woman. Yājñavalkya's praise of Maitreyī for her choice of knowledge over business suggests that he is somewhat surprised at her choice. Radhakrishnan's assessment of the meaning of Maitreyī's story with respect to the status of women, that 'this section indicates that the later subjection of women and their exclusion from Vedic studies do not have the support of the Upaniṣads',[49] is too general, too inattentive to the story in its move to generalize.

But—one may ask—how do we know all this? Well, Maitreyī's story in BU 2.4 and 4.5 is a concise story. It begins in *medias res* (middle), and does not explicitly tell us anything about Yājñavalkya's and Maitreyī's family life prior to Yājñavalkya's announced departure. The Upaniṣad does not even tell us that Yājñavalkya did not have a son. And yet, heeding the meaning of having no sons in traditional India, associated—among other things—with the thwarting of the expected relegation of wives to the protection and custody of her sons, we cannot resist the forceful suggestion embedded in the *Bṛhadāraṇyaka Upaniṣad* about Maitreyī's and Yājñavalkya's condition of childlessness. It is—obviously—the subtext of the entire story. It is the ground of the conflictual situation in Yājñavalkya's family as well as of the husband's and wife's personal crisis.

The latter part of the story is no more harmonious than the first. The story itself expresses Yājñavalkya's defensive, resentful, mood. As Maitreyī asks for his knowledge, he retorts: 'You are very dear to me; but you should concentrate!' Why does he say that? Maitreyī is a *brahmavādinī*; she seems eager to learn. Why does her husband tacitly rebuke her for being potentially inattentive? Why qualify his declaration of love with 'but'? Yājñavalkya seems to dislike the idea of instructing his wife. He seems

[49]*The Principal Upaniṣads*, edited by S. Radhakrishnan, George Allen & Unwin Ltd., 1953, p. 201.

reluctant to give her his knowledge. We may remember: he does not give her an offspring either.

Why, indeed, does not Yājñavalkya beget children with his two wives? It is difficult to answer such a question, although the Upanishadic story may have taken it for granted. If Yājñavalkya were impotent, or unwilling to spill his semen, he could have obtained children—if not for himself then for his wives—in quite a few other ways.[50] Yājñavalkya's resistance to Brahmanical offspring-immortality is presented in Maitreyī's story as deeply rooted. Actually, Yājñavalkya revolts against this aspect of Vedic culture and values; and in the course of his revolt he deprives both his wives of this channel of immortality. Moreover, according to the artful and subtle presentation of their relationship by the *Bṛhadāraṇyaka Upaniṣad*, Yājñavalkya can hardly face his wife's frustration at being deprived of an offspring ('immortality'). As seen above, at a certain point in the course of her husband's incessant talk, Maitreyī says: 'You have confused me!' (*atraiva mā bhagavān amūmuhāt*). And she says precisely what the context of her confusion (frustration) is; it is Yājñavalkya's assertion that there is no awareness (*saṃjñā*) after death. Thus, Maitreyī simply says that Yājñavalkya has not delivered what he had apparently promised (to provide Maitreyī with a teaching that would provide an alternative immortality). For indeed, if there is no consciousness (and therefore no individuality) after death, where is *my* promised immortality? Maitreyī seems to wonder. What then is Yājñavalkya's response? He denies his failure, putting the blame on Maitreyī: '(What I have said so far), my lady, is enough for understanding!' (*alaṁ vā ara idaṁ vijñānāya*). In other words, it is your problem, my dear wife, that you do not understand. I have said enough! The abruptness of Yājñavalkya's *alaṃ*-discourse, together with his initial admonition to Maitreyī to concentrate on his qualified declaration of love, frame this exchange in a hermeneutically interesting way that has been consistently unnoticed.

We can assess by now how inaccurate is the picture drawn by scholars of the Yājñavalkya couple united by harmony, spirituality, and love. Obviously, the person and suffering of Maitreyī are grossly overlooked. R.D. Ranade sees Maitreyī as 'the spiritual wife of Yājñavalkya',[51] his 'fit consort'. About Yājñavalkya and Maitreyī, he says:

[50]See in this respect Manu's description of the different 'types of son' (*Manusmṛti* 9.165–179).

[51] R.D. Ranade, *A Constructive Survey of Upanishadic Philosophy*, Oriental Book Agency, 1926, p. 383.

An irascible philosopher by nature ... he (Yājñavalkya) seems nevertheless to possess the kindness of human feelings, especially in his relations with his wife Maitreyī. Given to bigamy, he nevertheless maintains a strict spiritual relation with Maitreyī, while Kātyāyanī, his other wife, he regards merely as a woman of the world and prizes accordingly.[52]

Within this story of conflict and desertion, Yājñavalkya comes forth with one of the most famous teachings in the Upanishadic literature, repeated in BU 2.4.5 and 4.5.6: *sa hovāca na vā are patyuḥ kāmāya patiḥpriyo bhavaty ātmanas tu kāmāya patiḥ priyo bhavati*. The husband is not dear (or does not become dear) for the love for the husband, but rather, the husband becomes dear for the love for the *ātman*. Then he says: 'The wife is not dear for the love for the wife ...' and 'a son is not dear for the love for the son ...', towards the end of a long sequence. Thus, Yājñavalkya starts with love for husbands, moves on to love for wives, sons, riches, cattle, brahmanhood, kshatrahood, the worlds, gods, Vedas, beings, everything. The sequence is interesting; Yājñavalkya starts by addressing the immediate context of his clash with Maitreyī, and then moves on to other things, making the resolution of his conflict with his wife—as well as his own inner conflict—into an occasion for thinking. From the perspective of our margin, this statement is somewhat enigmatic; does it express a matter of fact? Does it recommend value? Does it suggest both? Can the storytelling (Yājñavalkya's encounter with Maitreyī) be of any help? P. Olivelle's translation takes Yājñavalkya's intention to be factual. Yājñavalkya seems to say, according to Olivelle's translation—literally almost accurate[53]—that as a matter of fact, when one (a wife) loves a husband one actually loves herself: 'One holds a husband dear, you see, not out of love for the husband; rather, it is out of love for oneself (*ātman*) that one holds a husband dear.'[54] H. Oldenberg endorses a similar understanding: 'We love, Yājñavalkya says, really only our self; all other love flows from this love and serves it.'[55] R.D. Ranade—located differently on the margin—senses danger here and, somewhat enigmatically combining fact and value, warns:

It is important to remember that this passage is not to be interpreted in the interest of an egoistic theory of morals, as some have done, but only in the interest

[52]Ibid., p. 19.

[53]The original Sanskrit contains two consecutive datives ... *kāmāya patyuḥ* ...' for love for husband.' Olivelle takes the first dative in the ablative sense ('out of love ...')

[54]*Upaniṣads*, trans. by P. Olivelle, Oxford University Press, 1996, p. 69.

[55]H. Oldenberg, *The Doctrine of the Upaniṣads and the Early Buddhism*, trans. by S.B. Shrotri, Motilal Banarsidass, 1991, p. 115.

of the theory of Self-Realization. We have not to understand that the wife or the husband or the sons are dear for one's own sake, interpreting the word Ātman in an egoistic sense. The word Ātman which comes at the end of the passage in the expression *Ātma vā are draṣṭavyaḥ* forbids an egoistic interpretation of that word in the previous sentences. We are thus obliged to interpret the word Ātman throughout the passage in the sense of the Self proper, the Ultimate Reality, and, therefore, to understand that the love that we bear to the wife or the husband or the sons is only an aspect of, or a reflection of, the love that we bear to the Self. It is, in fact, for the sake of the Self that all these things become dear to us. This Self the Bṛhadāraṇyaka enjoins upon us to realize by means of contemplation.[56]

Many commentators and translators stumble here, and not only at the double dative. M. Müller is strikingly less accurate than usual, and also most enigmatic: 'Verily, a husband is not dear, that you may love the husband; but that you may love the Self, therefore a husband is dear.'[57] Missing—unusually—in literal accuracy is also S. Radhakrishnan's translation; he overlooks the noun 'love' (*kāma*) in the dative case: 'Verily, not for the sake of the husband is the husband dear but for the sake of the Self is the husband dear.'[58] R.E. Hume is accurate as usual: 'Lo, verily, not for love of the husband is a husband dear, but for love of the Soul (*Ātman*) a husband is dear.'[59] R.D. Ranade makes use of the expression 'for the sake of' to suggest a sense of moral recommendation: 'It is not for the sake of the husband, that a husband is dear, but for the sake of the *Ātman*.'[60]

However similar or different from each other, the translations of BU 2.4.5 and 4.5.6 are somewhat ambiguous, apparently expressing the presentation of Yājñavalkya's uncertainty while groping for the resolution of tension with his wife. The self (or Self) is very important, while all the entities such as husband, wife, sons, wealth, etc., are less important. However, the BU 2.4.5 (and 4.5.6) repeat the same paradigm twelve times without resolving the conflict; Yājñavalkya's terminology strongly suggests its initial connection of the situation with his wife's challenge. Husband, wife, son, wealth; these indeed are the formative concepts

[56]R.D. Ranade, A *Constructive Survey of Upanishadic Philosophy*, Oriental Book Agency, 1926, p. 304.

[57]*The Upaniṣads*, trans. by Max Müller, rep. Dover Publications, 1962, p. 182.

[58]*The Principal Upaniṣads*, edited by S. Radhakrishnan, George Allen & Unwin Ltd., 1953, pp. 282–3.

[59]*The Thirteen Principal Upaniṣads*, trans. by R.E. Hume, Oxford University Press, 1921, p. 145.

[60]R.D. Ranade, A *Constructive Survey of Upanishadic Philosophy*, Oriental Book Agency, 1926, p. 303.

which comprise the definition of the situation at the beginning of the story.

Yājñavalkya ends his enigmatic series of examples with the clear recommendation: 'The self, my wife, should be seen, heard of, thought about and meditated upon' (*ātmā vā are draṣṭavyaḥ śrotavyo mantavyo nididhyāsitavyaḥ*). Yājñavalkya seems to yearn for a leap of consciousness, a leap which would end conflict and duality; in other words, a move to end tension, conflict, and inner as well as inter-subjective discord.

Yājñavalkya's eagerness to overcome duality is rooted in his fight with his wife. His teaching continues:

For where there is alleged duality, there one smells another (*yatra hi dvaitam iva bhavati tad itara itaraṃ jighrati*), there one sees another, hears another, speaks to another, thinks of another, understands another. But where everything becomes the self then whom can one smell and by what (*yatra tvasya sarvam ātmaivabhūt, tat kena kaṃ jighret*)? Whom can one see and by what? Whom can one hear and by what? To whom can one speak and by what? Whom can one think of and by what? Whom can one understand and by what?

And Yājñavalkya, apparently filled with the great joy of discovery, concludes: *yenedaṃ sarvaṃ vijānati, taṃ kena vijanīyāt, vijñātāram are kena vijānīyād iti*. 'How can one know that by which one knows? How (by what means), my wife, can one know the knower?'

Yājñavalkya's speech is, indeed, a 'culminating point' in the Upanishadic literature. Emerging out of his forthcoming separation from his wife, her insistent refusal to accept it, his own predicament over the renunciation of sons and Vedic immortality, Yājñavalkya moves away from life-in-the world—based on duality—to a different existence of unity and deadly immortality. Human activities such as smelling, hearing, speaking, seeing, understanding, etc., do involve a condition of duality in which separation from wife and life is hardly acceptable. Thus duality involves abysmal, unbearable pain for Yājñavalkya, and he wishes to transcend it altogether. However, the other's disturbing presence is not easily effaced. Maitreyī seems adamant in her expression of frustration and incomprehension.

Heeding Maitreyī's frustration and refusal to accept spiritual unity-in-separation offered her by Yājñavalkya, there is a remarkable note of ambiguity in the Upanishadic voice. The conflict is not fully resolved; and since rooted in unresolved conflict, ambiguity betokens a measure of ambivalence as well.

Indeed, one may wonder again about Yājñavalkya's condition. Had he 'known' the sublime truth of the unity of all *before* being challenged by

Maitreyī? Is he the unchanging, solid, secure, unmoved sage? His speech, culminating in the articulation of unity against the experience of duality, vibrates with freshness, a sense of discovery. The story opens with his decision to move into the wilderness. It may signify that he too is not whole, plagued with dissatisfaction, seeking something not to be found in the family and village.

On the philosopher's margin, however, scholars ignore the narrative context and consider instead in this context the question of Yājñavalkya's compatibility with Kant. Thus H. Oldenberg observes:

> Then it is true, Yājñavalkya continues, when he goes from loving of self over to the thinking of self, 'One should verily see, hear, think, absorb oneself in it, O Maitreyī; through seeing, hearing, knowing the Self, verily, is all this (existence) known.' Even in these sentences, it would be gratuitous to find agreement with Kant.[61]

And Oldenberg seems to sigh deeply over the results of his comparison of Yājñavalkya ('the sage') with the ideas of the critique of pure reason:

> We have already discussed this dialogue above. The sage wants to say that all values are in the Self; all the wealth of life emanates from the Self; thus all knowledge is included in the knowledge of the Self; an idea being, so to say, on another level than the posing of the problem of epistemology, as the critique of pure reason.[62]

Similarly ignoring the narrative context, A.B. Keith dislikes Yājñavalkya's metaphysics; he blames Indian philosophy for perceived character faults of the Indians, and in Yājñavalkya's personal case, Keith thinks that his metaphysical ideas 'drove Yājñavalkya to abandon his dear wife forever, and not, as in the more sympathetic version of the Buddha legend, subject to the possibility of relations of friendship at a later period'.[63] He sums up the main ideas of Yājñavalkya in the Maitreyī discourse as follows:

> The essence of its doctrine is that of the saving of the individual soul, which must resolutely refuse to allow itself to be blinded by the ordinary desires and sympathies of mankind. This is a logical conclusion of the metaphysics of the system. When Yājñavalkya declares that the husband is dear to the wife not for his own sake but

[61]H. Oldenberg, *The Doctrine of the Upaniṣads and the Early Buddhism*, trans. by S.B. Shrotri, Motilal Banarsidass, 1991, p. 115.

[62]Ibid.

[63]A.B. Keith, *The Religion and Philosophy of the Veda and Upanishads*, Harvard University Press, 1925, rep. Motilal Banarsidass, 1989.

for the sake of the self, and applies the same principle to the other relations of human life, to the wife, to the children, to riches, to other men, to the gods, and the universe, he is doubtless concerned mainly with a metaphysical doctrine; but not only is the selfish conclusion obvious to ordinary minds, but it is essentially connected with the teaching itself, for on ultimate analysis the aim of the self turns out to be the annihilation of every human desire and activity, an ideal which renders all active philanthropy idle, and which has caused the chief virtues of India to take the form of resignation, passive compassion, and charity.[64]

Keith is somewhat too harsh in his judgement of Yājñavalkya's theory ('foolish conclusion') and ('selfish') character; he ignores the context drawn by the Upanishadic story, a context of the deepest personal and marital conflict and distress. The picture emerging from the story invites some sympathy for Yājñavalkya, since he seems to cope with a terribly real problem (death). He disdains, as seen earlier, the idea of Vedic immortality consisting of offspring; he revolts against offspring-immortality and decides on leaving home. But he is caught in a most bitter conflict with his perspicacious wife, who exposes his real failure to provide for an older ('normal') type of immortality while groping for and failing to articulate a new type of immortality. What can a husband say to his wife whose deepest desire (for offspring) has been frustrated by his own (not) doing? What can a man say to himself in his failure to pay the debt to his ancestors (say, his father)? In the Upanishadic crucible comprising a relentlessly inquisitive wife who faces long years of solitude and dissatisfaction, he does find a solution. Immersed in the process of thinking hard, he is excited and forcefully pushes Maitreyī (and the reader) to let him go. But questions remain: Who goes to the wilderness? Why not stay at home, clothed in the new immortality? Why not wait for Maitreyī's spark of understanding? And what about Kātyāyanī?

We suggest that Maitreyī's story is one occasion of *ātman*-theory formation (or discovery), and the Upanishadic voice reflects—with a measure of ambivalence—on the situation within which Yājñavalkya's sublime philosophy emerges. Yājñavalkya legitimizes and justifies his fateful separation from Maitreyī (and Kātyāyanī). Like so many men, he wants her (and possibly himself) to *understand* (please, consider—reflect, heed, think deeply—*nididhyāsasva*), and in his endeavour to convince her (and himself) of the viability of a new immortality, he comes forth with the Upanishadic metaphysics of non-duality. He says: 'Let me go, for your love (desire) for me is misdirected, as the husband becomes dear

[64]Ibid.

not out of the desire for the husband but for the love for *ātman*.' And similarly with respect to wife and children: 'My own love for you and my yearning and dependence on children is nourished and made possible by the ultimate reference (*ātman*).' 'It is illusory,' Yājñavalkya seems to say, 'to desire a husband (like me) for my own sake. Let me go!' Yājñavalkya desparately—and successfully as well—reaches a solution in his search for immortality independent of offspring. He analyses the condition of desire and finds refreshing, guilt-free enlightenment ('knowledge'). Conflict and distress disappear in a universe of unity.

If accepted and understood by Maitreyī, such a beautiful insight could alleviate her burden of childlessness and mortality. However, none of the two versions of her story offer such a happy ending; BU 2.4 does not end the story at all (no response of Maitreyī or reference to Yājñavalkya's departure); BU 4.5.15 ends with Yājñavalkya's leaving for the wilderness with no reference to Maitreyī's state of mind (after her previous protest of confusion, one would expect such a reference to her). According to BU 4.5.15, Yājñavalkya ends his speech about unity and the One free of suffering and injury (*na vyathate na riṣyati*) with the words: 'Thus you have been instructed, Maitreyī; this is, indeed, immortality, my wife' (*iti uktānuśāsanāsi maitreyī; etāvad are khalvamṛtatvam*). Saying these words, Yājñavalkya leaves for the wilderness (*iti hoktvā yājñavalkyo vijahāra*).

As mentioned above, there is no response on Maitreyī's part. This eloquent silence resonates beyond the story's end. Maitreyī's reconciled farewell to her husband at the end of the Maitreyī episode (in both versions, BU 2.4 and 4.5) is conspicuously absent. Deserted by her husband, she remains childless, hopeless, and mortal in every way.

FOUR

Under-Reading Multiple Vocality: The Case of the Good Boy and the Angry Father

Death is apparently a weakness, often undesired. And yet, though closely associated with inferiority, weakness is not inferiority. The latter implies openness, a potential for and mode of awakening to one's weakness (such as mortality), and an awareness of something different, namely, excellence (such as immortality). Mortality is not exempt from the list of weaknesses-turned-inferiorities described in the Upaniṣads. Some of the more well-known Upanishadic stories focus on this infirmity of the person. In one of them, the story of Naciketas told in the *Kaṭha Upaniṣad*, the King of Death (Yama) tells the boy of the nature of the self, free of death.

The wise one—

he is not born, he does not die;
he had not come from anywhere;
he has not become anyone.

He is unborn and eternal, primeval and everlasting.
And he is not killed, when the body is killed.[1]

These references to the nature of the wise one—who knows Brahman and the syllable Om—may easily be abstracted from the story of the boy who experiences death. Such references have become familiar to any reader of Indian scriptures. The phrases quoted above sound particularly familiar since they are almost identical with the references to the *ātman*

[1] *The Early Upaniṣads*, annotated text and trans. by P. Olivelle, Oxford University Press, 1998, p. 385. P. Olivelle notes, that at this point 'the dialogue between Naciketas and Death appears to end'.

in the Bhagavadgītā 2.20. He is not born, neither does he ever die (*na jāyate mriyate vā kadācin*); it has never come into being, and neither will it ever come to be again; it is unborn (*aja*), eternal, everlasting, and ancient (*ajo nityaḥ śaśvato 'yaṃ purāṇaḥ*); it is not killed when the body is killed. However, familiarity may deceive if interpreted as pointing to the abstract, context-free nature of the 'knowledge' involved. In both cases—the *Kaṭha Upaniṣad* and the Bhagavadgītā—the heroes' awakening to knowledge is forcefully embedded in contexts of inferiority and crisis, an integral part of powerful and significant narratives.

Then, after Naciketas received this body of knowledge,
And the entire set of yogic rules taught by death,
He attained *brahman*, he became free from aging and death;
So will others who know this teaching about the self.[2]

The narrative refers to a heated, dialogical space wherein knowledge is transmitted and shared. The Upanishadic narrative the *Kaṭha Upaniṣad* suggests, should not be forgotten; the experiential potential of storytelling is part of the value and indispensability of the Upanishadic narrative:

The wise man who hears or tells
The tale of Naciketas,
An ancient tale told by Death,
Will rejoice in brahman's world.[3]

What happens in the encounter with the enigmatic, hot, secret, and contextual sayings contained in the Upaniṣads? Life experiences and satisfactions vary; inner space sets different constraints. Of many kinds are the onlookers peering at the Upanishadic crucible; 'value-free' scientists, orientalists, reformers, seekers, poets, Westerners, neo-Hindus, etc. Often what they see becomes what we see in a process of textual reception and interpretation that incrementally cannibalizes its origin. Through a process of under-reading and decontextualization, interpretations of the Upanishadic situation, context, and metaphysics become banalized on the margin. This book attempts both local remedy, rereading several Upanishadic stories frequently lost in pursuit of sayings, and a more profound corrective, a hermeneutics based on attentive textual and contextual reading.

We have seen earlier how Maitreyī's voice has been traditionally

[2]Ibid., p. 403.
[3]Ibid., p. 391.

avoided, subdued, suppressed, conflated. Her voice of bitter challenge, protest, and yearning for 'immortality' has invariably been conflated with her husband's and the narrator's. No boundary between the deserted, childless woman's cry and her husband's (individualistic, spiritual) voice has been established. The result is loss of essential otherness, and a mode of reading which involves diminished contact with the text:

> A sensing and the object sensed, an intention and its realization, one person and another, are confluent when there is no appreciation of a boundary between them, *when there is no discrimination of the points of difference or otherness that distinguishes them*. Without this sense of boundary—this sense of *something other* to be noticed, approached, manipulated, enjoyed—there can be no emergence and development of the figure/ground, hence no awareness, hence no excitement, hence no contact.[4]

Indeed, Maitreyī's story *is* exciting. Moreover, Yājñavalkya's verbalization within the unfolding crisis with his wife is possibly one of the 'culminating points in the Upaniṣads'. However, vocal (and personal) multiplicity must be admitted in order for the story and the context to emerge.

Or let us listen again to Satyakāma Jābāla's voice, as he quotes—for his prospective master—his mother's 'I do not know your lineage, son'. Satyakāma does not paraphrase his mother's tale, but repeats it *verbatim*, in the first person: 'In my youth I was a maiden roaming about in many places ....' Even in this case, of course, the following question is justified: Is Satyakāma's voice the same as his mother's? Set in the context of Satyakāma's encounter with his guru, his voice is different. The recognition of the boundary between the boy's voice and his mother's—though seemingly the same—helps in the contact with Upanishadic otherness.

The old Indian tradition itself seems to have reflected on this issue of otherness-digestion. Thus, according to the famous story from the *Śatapathabrāhmaṇa*, the two Aśvins cut off Dadhyañc's head, replacing it with a horse-head (like their own).[5] Facing the scorching, unfathomable language embodying the great secrets of creation, sacrifice, and completion, these Aśvins—doctors, surgeons of the gods—have to tame and banalize language for their own understanding; thus they cook, compromise and subdue Dadhyañc's otherness. Under certain circumstances, it seems, this excruciating surgery of banalization is (at least temporally and locally) essential; however, unchecked, a danger of triviality creeps in, and clarity

[4]Perls, Hefferline, and Goodman, *Gestalt Therapy*, p. 153.

[5]See *Śatapathabrāhmaṇa*, trans. by J. Eggeling, rep. by Motilal Banarsidass, 1978, 14.1.

obtained as otherness fades away. A subtle balance is needed whereby the other is allowed any available inner space, but is also made somewhat banal and understood. This is what the translators of the Upaniṣads experience as the tension between 'clarity' and 'accuracy'.[6]

Among the numerous contexts and points of views, one is particularly valuable—the Upanishadic meta-testimony itself, in the form of storytelling. As testimony and meta-testimony are integrated, the most powerful equivalent—in the textual domain—to real experience is produced. The story weaves emotion, meanings, aspirations, events, relationships, culture, tradition, economy, politics, and wisdom. It is the story in all its detail, as an organic whole, which is closest to 'experience'.

The relationship of Yājñavalkya and his wife Maitreyī, Upakosala's dejection focused on his inner condition of 'multiple desires', Satyakāma's doubts about his 'identity', Śvetaketu's pride and individuality, Naciketas' ambiguous challenge to his father during a fateful sacrifice, etc., are all made known to us through articulated, sophisticated references to integrated fields of emotion, values, tradition, social tensions, individuality, etc. An entire universe of multidimensional quality unfolds in the Upanishadic story. As a translation of such a universe, narrative detail is invaluable for a richer approach to the Upanishadic experience. Thus, the fact that Vājaśravas' cows are described as lean and feeble by Vājaśravas' son (Naciketas)—and not by the narrator—invites interpretation. Commentators and scholars may overlook the distinction between narrative voice and the voices of characters within the narrative, a partial-reading approach which reduces openness to the very universe of thought and experience unfolded and revealed through the Upanishadic story. This tendentious reading strategy may express a sense of impatient arrogance; it may also suggest an impatient seeking, a shucking of a (mis)perceived shell (the story) in search of a core truth (the saying).

Some margin-dwellers wish to understand India or the Upanishadic culture as means for other ends. They do not pay much attention to the details of the stories, to the different voices heard in the narrative, or to the individuality of the protagonists of Indian wisdom. Among these under-readers, for example, are scientific 'reformers' such as R. Lannoy, who regards the objective social setting so important that he considers the Upanishadic testimony superfluous. Although 'Upaniṣads are the vedic scriptures *par excellence* of Hinduism', as Olivelle says,[7] Lannoy

[6]*Upaniṣads*, trans. by P. Olivelle, Oxford University Press, 1996, p. xxxiii.
[7]Ibid., p. xxiii.

does not take these texts too seriously in his lengthy study, *The Speaking Tree*, although he makes many observations on pertinent aspects of the Upanishadic lifestyle and meanings:

> The perfected sage endures a period of regression, of initiatory latency separate from the tribe, but returns to play a role in the collective; in a sense he is more indissolubly a part of it, since he has been in a state of unified awareness. His role is within the society, as the seer who *sees* the interconnectedness of all things. Vedic sacrificial rites are a dramatization of integral oneness; there is no question of the sage *renouncing* society. This ideal of cosmic interconnectedness is never abandoned in Indian history, but with the disintegration of the sacrificial, tribal, potlatch-type society and its replacement by the caste society, an entirely new emphasis is given to the Zero idea and the Zero ideal. The state of spiritual revelation when the opposites are transcended, variously called *nirvāṇa*, *mokṣa*, *mukti*, or *samādhi*, can equally be conceived as absolute plenitude or absolute negation.[8]

This secondary verbalization—based on closer and better conceptualizations of India—of Lannoy's is not wrong; moreover, it contains suggestions of important ideas central to Vedic orientation and self-understanding (the interconnectedness of all things). However, the import of passages like this is abstract, distant, and—generally—less interesting than the stories to which it gives such cursory attention, for Lannoy is not truly seeking an encounter with the other. Indeed, Lannoy is explicit and sincere in exposing his particular position on the margin from the very beginning of the Introduction:

> This book is an analytical study of Indian culture and society, with the chief aim of identifying the origins of the nation's contemporary problems. Though generally manifested in urgent economic terms, these have their root causes in the historical development of India's system of values and thought, as reflected in its cultural and social organization.[9]

Lannoy seems determined to assess India's capacity for change in the present, and he subjugates his research and thinking to this goal:

> The complexity of India's task is on a scale unprecedented for any previous period in its history, and can be accomplished only if its people themselves have the capacity for change—and the will to take decisive action.[10]

[8]R. Lannoy, *The Speaking Tree*, Oxford University Press, 1971, p. 346.
[9]Ibid., p. xv.
[10]Ibid.

As this glimpse of Lannoy at work shows, the reformer is one of the margin-dwellers least attentive to the subjectivity and otherness of the Indian heroes. His inner space—reform-ridden—is particularly incompatible with the obscure, the contextual and the secret; the Upanishadic abyss, from which arise intimate whispers as if meant to be unheard or misunderstood, is particularly difficult for outsiders to understand. How foreign and bizarre is the goose speaking to Satyakāma Jābāla;[11] who is this goose? Who is the bull, and the fire? Are their identity and nature connected with their teaching? In this darkness, one seems to need all the help one can get.

Fortunately, the Upanishadic margins are densely populated with the best minds attracted to the crucible of transformation, understanding, and—possibly—alleviation of sorrow. This crowd strains its eyes and minds gazing at fathers and sons, husbands and wives, teachers and disciples, inimical and friendly sages, who exchange precious communications touching on terribly important secrets of death, life, selfhood, and immortality.

Locations on the Upanishadic margins are distinguished by the different contexts—and levels of banalization—they set and use for interpretation of the Upanishadic situation, experience, and wisdom. We have sketched the reformer's location, a particularly unpromising one. Edward Said's anti-essentialist hermeneutics, while not comfortably applicable to the domain of Indian spirituality and scholarship, are instructive with respect to the reformer's location:

> Everyone who writes about the Orient must locate himself vis-à-vis the Orient; translated into his text, this location includes the kind of narrative voice he adopts, the type of structure he builds, the kind of images, themes, motifs that circulate in his text—all of which add up to deliberate ways of addressing the reader, containing the Orient, and finally, representing it or speaking on its behalf.[12]

'Orientalism' as described by Said is only one of the sources of under-reading Indian otherness, one that aptly situates the reforming consciousness. Yet much of the scholarly attention to the otherness of India and its mystic-philosophical traditions is largely free of orientalism. One type of such invaluable scholarship on the margin is the scientific exploration of the conditions and background from which the Upanishadic whispers and testimony have emerged. Much of today's Indology consists of scientific, 'value-free' efforts at seeing India as an object. Such efforts are nonetheless constructive, providing vital information about the context

[11]See ChU 4.6.

[12]E.W. Said, *Orientalism*, Vintage Books, 1978, p. 20.

of the Upanishadic culture and 'way of life'. W. Halbfass' description of the great sociologist Max Weber's orientation towards India succinctly describes this 'scientific' hermeneutic:

India was highly significant for Weber's comparative enterprise of European self-understanding. But it was not a potential supplier of meaning and values. India itself, like other traditions, was an object of 'value-free' research and rational understanding.[13]

Here too, on the margin populated with 'value-free' scientists (exploring India as an 'object'), location is most variable. Scientists and cool historians banalize the Upanishadic testimony in most useful ways, providing a meta-context for these expressions of the spirit of ancient India. P. Olivelle is one such cool, sympathetic historian. His introduction to his translation of the Upaniṣads expresses his deep appreciation for these 'documents' along with the self-perception of a scholar who views the Upanishadic existence from without, objectifying it:

The Upaniṣads translated here represent some of the most important literary products in the history of Indian culture and religion, both because they played a critical role in the development of religious ideas in India and because they are valuable as sources for our understanding of the religious, social, and intellectual history of ancient India.[14]

Firmly grounded in the historian's site and watchful perspective, Olivelle is duly cautious; he does not address the validity and religious significance of the Upaniṣads.[15] Olivelle's key words are 'information' and 'background'. The ideology of the impartial historian is made explicit by Olivelle thus:

In this introduction, I want to supplement and to organize the information scattered in the notes—information that will give the reader the social, religious, and cultural background of these documents, information without which much of what is said in them cannot be understood.[16]

Indeed, information about the times, economy, tradition, past, religious beliefs, etc., is valuable in approaching the Upanishadic crucible.

The sixth or seventh century BC—the Upanishadic age—is characterized

[13]W. Halbfass, *India and Europe*, SUNY, 1988, p. 144.

[14]*Upaniṣads*, trans. by P. Olivelle, Oxford University Press, 1996, p. xxiii.

[15]See the discussion (Ch. 2) on Keith's discourse about the meaninglessness of *Tat tvam asi*.

[16]*Upaniṣads*, trans. by P. Olivelle, Oxford University Press, 1996, p. xxiii.

by scholars as an epoch of surplus of food, urbanization and the 'rise of individualism'[17] as well as of 'spiritual malaise'.[18] There emerged 'ideologies (that) shared the view that human life was essentially suffering, which even death could not end; death is a mere interlude in the never-ending cycle of rebirth'.[19] Some scholars raise the hypothesis that 'many unknown diseases and epidemics may have erupted, causing widespread death. Gombrich (1988, 58–59) was the first to suggest a relation between such urban epidemics and the ascetic ideologies of ancient India.'[20] 'The new religious ideologies and the increasingly widespread ascetic life-styles fostered by urbanization stood in sharp contrast to the Vedic religious world centred around the householder and his duties of sacrifice and procreation.'[21] These may have been the circumstances in which the Upanishadic experience and storytelling emerged and, as such, can usefully inform our reading of the stories, without, however, substituting for them.

Olivelle is right about the potential contribution of meta-context (social, religious, economic, etc.) to our understanding of the—often obscure—Upanishadic 'documents'. He indirectly reflects on the enormous difficulty remaining in understanding the documents thus lit by meta-contextual illumination. This difficulty is particularly acute in the case of translating the Upaniṣads: 'The ideal of every translator is clarity and accuracy, two goals that often tend to exclude each other.'[22] Since 'accuracy' betokens touching and giving expression to the other's innermost otherness ('intentions'), it contradicts the move towards lucidity (clarity), the inevitable product of exegetic banalization. Indeed, claims for value-free translation and comprehension are particularly commensurate with the objectification of the other, and thus with the location of the cool historian on the Upanishadic margin:

> In this introduction I have avoided speaking of the 'philosophy of the Upaniṣads,' a common feature of most introductions to their translations. These documents were composed over several centuries and in various regions, and it is futile to try to discover a single doctrine or philosophy in them. Different theologians,

[17]See P. Olivelle, *The Sannyāsa Upaniṣads*, Oxford University Press, 1992, p. 32: 'Cities, kings, and the merchant class contributed to the rise of individualism ...'

[18]Ibid., p. 34.

[19]Ibid., p. 33.

[20]Ibid., p. 35. 'The widespread death from previously unknown diseases might have provided a catalyst for considering life as bondage and suffering' (p. 35).

[21]Ibid.

[22]Ibid., p. xiii.

philosophers, and pious readers down the centuries both in India and abroad have discovered different 'truths' and 'philosophies' in them. That has been, after all, the common fate of scriptures in all religions. Even in the future, that is an enterprise best left to the readers themselves, and the prudent translator will try and step aside and not get in their way.[23]

Olivelle's construct of a 'prudent translator' capable of 'step[ping] aside' and providing readers with unmediated textual access obscures the inevitable mediations of meaning inherent in the act of translation. Objectifying language, viewing words as objects of exchange rather than vitally slippery moments of experience themselves, invites the kind of under-reading that Upanishadic stories have been put through.

The essential logic of the Upanishadic experience and storytelling is the compatibility (commensurability) of inferiority and knowledge, of disease and medicine. One must be sick in order to be cured. A person heeding himself whole or healthy would not go for remedy. The Upanishadic narrative invariably refers to and describes the condition of various modes of inferiority ('disease'), combining such descriptions (narratives) with the emergence of knowledge. The infirmities listed as such in the culture of the Upaniṣads are many; householdership impervious to the call of renunciation and the absolute, mortality, pride, desire, scholarship devoid of knowledge of self, inner doubt concerning entitlement, and others. However variable the states of disease, remedial knowledge articulated as the unity of all, or the ultimate reality of the one pure subject is uniform and one. Thus, the uniformity of remedial knowledge suggests the possibility of de-contextualizing the Upanishadic narrative to the effect that knowledge can be independent of the circumstances of its emergence or transmission. However, from the evident uniformity of remedial knowledge does not follow its independence of context. For there is a middle way; a bunch of inferiorities is cured by potent, uniform Upanishadic knowledge. The contexts of knowledge-discovery may vary, yet they are always there, inherently commensurate with remedial knowledge.

As a unified entity, integrating many aspects and meanings of profound, almost untouchable existence, the Upanishadic story is inevitably under-read. Exploring this point of under-reading much-read stories, we follow here one of the better known Upanishadic stories, the story of Naciketas and Vājaśravas, his father, told in the *Kaṭha Upaniṣad*, one much dealt

[23]Ibid., p. xxiv.

with in Upanishadic exegesis.[24] The *Kaṭha Upaniṣad* is probably more widely known than any other Upaniṣad. It formed part of the Persian translation, was rendered into English by Rammohan Roy, and has since been frequently quoted by English, French, and German writers as one of the most perfect specimens of the mystic philosophy and poetry of the ancient Hindus.[25]

The Naciketas story has its roots in the most ancient history of textual Hinduism. Sāyana suggests that the *Ṛg-Veda* refers to Naciketas in RV 10.135, and a fuller version of the story is found in *Taittirīya Brāhmaṇa* (3.1.8):

> Vājaśravas, wishing for rewards, sacrificed all his wealth. He had a son, called Naciketas. While he was still a boy, faith entered into him at the time when the cows that were to be given (by his father) as presents to the priests, were brought in. He said: 'Father, to whom wilt thou give me?' He said so a second time and third time. The father turned round and said to him: 'To Death, I give thee.'[26]

The scorching utterances in the midst of the Upanishadic crucible are apprehended in accordance with inner space; everyone is endowed with sufficient inner space to sense the father's anger, thus we understand 'the father is angry'. Moreover, in this case we follow the great Śaṅkara; the father is angry (*kruddha*), he says. All translators and commentators seem to duplicate and repeat the same inner space in this context. R.E. Hume observes: 'Thereupon the father, in anger at the veiled reproof, exclaims: "Oh! Go to Hades!"'[27] According to M. Müller: 'He said it a second and a third time. Then the father replied (angrily): "I shall give thee unto Death."'[28] P. Olivelle says: 'His father yelled at him: "I'll give you to Death!"'[29] S. Radhakrishnan's version is: 'When he persisted in his request, his father in rage said: "Unto Yama, I give thee."'[30]

Emotion is a very powerful means of orientation. If a father is angry at his son, and if anger is perceived as an inexplicable fact, this is potentially trivial; all fathers and all people know this anger. Yet, finding more available

[24]For a fascinating look at this story in the context of parental sacrifice and divine appetite, see David Shulman, *The Hungry God*, University of Chicago Press, 1993.

[25]*The Upaniṣads*, trans. by Max Müller, rep. Dover Publications, 1962, p. xxi.

[26]Ibid., p. xxiii.

[27]*The Thirteen Principal Upaniṣads*, trans. by R.E. Hume, Oxford University Press, 1921, p. 341.

[28]*The Upaniṣads*, trans. by Max Müller, rep. Dover Publications, 1962, p. 2.

[29]*Upaniṣads*, trans. by P. Olivelle, Oxford University Press, 1921, p. 232.

[30]*The Principal Upaniṣads*, edited by S. Radhakrishnan, Georges Allen & Unwin Ltd., p. 593.

space, one may ask: why is the father angry? The answer seems obvious: the son asks a seemingly identical question—'to whom will you give me?'—*three times in a row*. The father is apparently unwilling (or unable) to answer the question the first time. Thus the second questioning is somewhat irritating, and the third one even more so; the father is angry and says: To death I shall give you! (*mṛtyave tvā dadāmi*).

But emotion is obviously insufficient. The Upanishadic situation contains many other components in need of attention. A father performs a sacrifice. He—apparently—gives everything, staking himself through his cows.[31] His son, as if possessed by a rush of 'energy' (*śraddhā*) from without (the energy enters—*āviveśa*—him), questions: 'Father, to whom will you give me?' (*tāta, kasmai māṃ dasyasi*). There is apparently no answer, so the boy asks again: 'Father, to whom will you give me?'; the father is again silent. The boy asks a third time: 'Father, to whom will you give me?' This time, the father articulates an answer: 'To death I shall give thee' (*mṛtyave tvā dadāmi*). The son goes to death (Yama). How does he do it? How does he die? Does his father kill him? Does he sacrifice him? These are questions suggested by the story, perhaps pertinent to its reading; yet they have not been asked by its readers. In general, themes inherent to the narrative have been overlooked in its interpretation.

Thus, at this point concerning the father's anger, scholars and commentators on the margin stop resisting premature understanding.[32] The father is angry—naturally—and sends his son to death (as punishment). Commentators do have more available space, but they do not let the otherness of the father's anger penetrate. They take the father's anger (and the irritating nature of the son's repetitive questions) for granted, in accordance with the most immediately available cognitive space.

And yet, the simplicity of the Upanishadic story is deceptive, as we have already witnessed. A spiritual wife seeks knowledge from her husband, as he intends to retreat to the forest.[33] This is natural since, after all, the wife is a 'fit consort' to her spiritual husband, and naturally interested in spiritual matters (*brahmavāda*). A boy returns home conceited; he thinks he knows much (having studied under a guru for twelve years). This is natural, since people full of knowledge can be conceited. A disciple is gloomy and tormented, a natural state since his guru has left him uninstructed. A boy

[31]See J.C. Heesterman, *The Broken World of Sacrifice*, University of Chicago Press, 1993.

[32]In one of our talks, D. Shulman referred to resistance as essential to 'real thinking' (referring to Wittgenstein in this context).

[33]See Maitreyī's story BU 2.4 and 4.5—discussed earlier, Ch. 3.

asks his mother who his father is. The mother, a good woman, tells the truth: 'I do not know'. The son believes her, and takes her answer intact to the guru, who accepts him as an apprentice since he did not deviate from the truth. A boy charged with 'faith' (*śraddhā*) challenges his father; the latter becomes angry and sends his son to the domain of the king of death (Yama). This is terribly unnatural. But on the margin even this story is substantially banalized, its dissonance too quickly resolved. The boy is good (wishing to help his father, to the point of offering himself as victim); the father is good (giving all his possessions in sacrifice); the king of death is also good (surrendering invaluable wisdom to the insistent boy). Everybody collaborates in the transmission of 'knowledge'. Yet, if a measure of resistance is invited, questions must be asked: Why is the father so angry (killing his son!)? Or is he perhaps not angry at all? And what precisely is the nature of the mysterious 'faith' (*śraddhā*) which enters the boy; how does its role in the narrative affect our conception of this 'faith'?

Though the Naciketas story is one of the most well-known stories in the Upaniṣads, major issues remain under-attended. One of them pertains to the value of the father's offering: Were his cows so bad-looking (and who says so)? This issue of Naciketas' perception of the cows—though apparently trivial—is connected to one of the central themes in the Upanishadic literature, the quest for immortality. Other unexplored paths follow questions about what happens to people after death. Secrecy and contextuality ground these questions in the Upanishadic literature. Some of the more important Upanishadic stories and theories are developed in contexts of deadly challenge, crisis, and death. Immortality and its variants are probably the most recurrent theme of the Upaniṣads. Maitreyī asks her husband about immortality. Arthabhaga asks Yājñavalkya about the fate of man after death. King Jaivali asks Śvetaketu questions about man's fate after death. Citra Gāṅgyāyani doubts whether Śvetaketu can place him—during sacrifice—in the right world, and tells his father, Uddālaka Āruṇi, about the two ways of man after death. The two oldest and classical Upaniṣads, the *Chāndogya* and the *Bṛhadāranyaka Upaniṣad* describe the process of death and the transition to immortality in several contexts. Obviously, the quest for immortality is a central preoccupation of this literature, and one that readers are invited to inhabit rather than resolve. The Naciketas story is no exception, and the tension between the old ('Vedic') and the new ('Upanishadic') immortality is part of the story and teaching. As in the case of the dialogue between Maitreyī and Yājñavalkya, the story should provide vital clues to the interpretation of the teaching.

Let us return to the lives of Naciketas and his father in their dramatic, fateful encounter told in the first chapter of the Kaṭhopaniṣad: A sacrifice is going on. A man called Uśan Vājaśravas performs an all-important sacrifice; it is perhaps the most significant event in his life as a householder. He—apparently—gives *everything* as gifts to the officiating brahmins, an exchange in the fateful game of life and death. Commentators and scholars do not identify with Vājaśravas' condition and its meanings. He—invariably—is perceived as 'pre-Upanishadic'. Heesterman provides a dramatic representation of the Vedic sacrifice:

> Sacrifice threatens to overwhelm the sacrificer, who now is at one with the victim. Like the cosmic man, the *puruṣa* of the cosmogonic hymn, he is himself at stake. In this way we can also understand that the Vedic ritual texts generally consider man as the *paśu*, the sacrificial animal *par excellence*, the one most fit for sacrifice because he alone among all *paśu* both sets the sacrifice in motion and undergoes it. ... Laying hands on the animal, the sacrificer becomes one with the animal, and so his vital breath is in danger of departing with the animal's when it is put to death. [34]

Exerting himself to the point of material and psychophysical exhaustion, Naciketas' father is involved in the quest for immortality. Indeed, he is on the verge of death; immortality is for him a tangible prospect and need.

Vājaśravas' son, Naciketas, is watching. Something happens to him. A sudden rush of 'religious energy' (*śraddhā*) enters (*āviveśa*) the boy. It is a crucial moment, the turning point in the story: 'As the sacrificial gifts were being led away, *śraddhā* entered him, though a boy' (*taṃ ha kumāraṃ santaṃ dakṣiṇāsu niyamānāsu śraddhāviveśa*). The cows given as gifts to the brahmins are—in Naciketas' eyes—lean and feeble, bad-looking. The storytelling expresses the change of voice underlying the perception of the cows. Naciketas thinks of the cows in a more poetic mood, expressing himself in the metre of *śloka* (rather than prose). Thus, the distinction between the narrator's and Naciketas' voice is clear-cut in the narrative. Paraphrases of the story often overlook this feature (multi-vocality) of the text.

This is how S. Radhakrishnan tells the story:

> A poor and pious *Brāhmaṇa*, Vājaśravas, performs a sacrifice and gives as presents to the priests a few old and feeble cows. His son, Naciketas, feeling disturbed by the unreality of his father's observance of the sacrifice, proposes that he himself

[34] J.C. Heesterman *The Broken World of Sacrifice*, University of Chicago Press, 1993, pp. 31–3.

may be offered as offering (*dakṣiṇā*) to a priest. When he persisted in his request, his father in rage said: 'Unto Yama, I give thee.' Naciketas goes to the abode of Yama and finding him absent, waits there for three days and nights unfed. Yama, on his return, offers three gifts in recompense for the delay and discomfort caused to Naciketas. For the first, Naciketas asked, 'Let me return alive to my father.' For the second, 'Tell me how my good works (*iṣṭa-pūrta*) may not be exhausted;' and for the third, 'Tell me the way to conquer re-death (*punar-mṛtyu*).'[35]

Thus, the distinction between the narrator's and Naciketas' voices disappears in Radhakrishnan's account; paraphrasing the story, he presents his own understanding of the narrative and situation.

Radhakrishnan is not the only reader who conflates the narrator's voice with Naciketas' in this context. K.L.S. Rao's rendering is typical: 'In the Kaṭha Vājaśravas performs a sacrifice in which he is supposed to give away the whole of his property, but is presented as giving away barren and useless cows....'[36] Rao's conflation of the narrator's and Naciketas' voices is compatible with his main argument concerning the meaning of *śraddhā* in the *Kaṭha Upaniṣad*:

> But when *śraddhā* entered into the heart of Naciketas, he was not filled with love or devotion to a god, but he reflected (*so 'manyata*) about the evil consequences of an ill-performed sacrifice. ... To sum up, then, the word *śraddhā* in the introductory part of Kaṭha retains its Brahmanic significance and does not carry its later Upanishadic connotation. It indicates an aspiration for heaven and a confidence in the efficacy of sacrifice as the appropriate means to realize it. ... It refers to a system of rites. In short, it is impersonal, ritualistic, external and formal; it is cold and legalistic. These characteristics, we maintain, are distinctive of the Brahmanic conception of *śraddhā*.

But what is the role of *śraddhā* in the narrative? In this context, Naciketas' perception of the cows is pertinent. Are the cows lean and feeble? Perhaps. But in the Upanishadic story, the diagnosis of the cows was made *by Naciketas* (not by the narrator), who reflects on the nature of the worlds and pleasures expected for one who gives such poor gifts to the brahmins.

The Kaṭha gives the description of the cows from Naciketas' perspective, in direct speech. There is a note of exaggeration in the extremity of the modifiers in his description:

[35]*The Principal Upaniṣads*, edited by S. Radhakrishnan, Georges Allen & Unwin Ltd., 1953 p. 593.

[36]K.L.S. Rao, *The Concept of Śraddhā*, Motilal Banarsidass, 1974, p. 81.

They've drunk *all* their water, eaten *all* their fodder,
They have been milked *dry*, they are *totally* barren—
'joyless' are those worlds called,
to which a man goes
who gives them as gifts.[37] [my italics]

Thus Naciketas—*śraddhā*-intoxicated, as it were—speaks *differently* in his observation of the cows and consequent gloomy (*ananda*) worlds his father is destined to go to. The metre changes from ordinary prose into the more poetic *śloka*:

Pītodakā Jagdhatṛṇā
Dugdhadohā nirindriyāḥ\
Ananda nāma te lokās
Tān sa gacchati tā dadat\\

This is obviously Naciketas' voice, not the narrator's.

Who are the particularly exhausted cows Naciketas reports seeing? Naciketas apparently sees *dying cows*, animals which have already drunk their water (*pitodaka*), eaten their grass (*jagdhatṛṇa*), given their milk (*dugdhadoha*) and are 'barren' (*nirindriya*). Śaṅkara interprets the description of the cows in the context of their adequacy as gifts given in the sacrifice (*pītodakā ityādinā dakṣiṇārtha gāvo viśeṣyante*). They are poor gifts, Śaṅkara suggests, since they have already drunk their water, etc.

We have wondered—resisting conflation, reflecting on the textual multi-vocality—whether the Upaniṣad wishes to describe in this context the 'real condition' of Naciketas' father's gift-cows, or whether Naciketas sees *through* the real cows—perhaps beautiful and healthy—into their essential mortality. It is somewhat unlikely that all Vājaśravas' cows were—simultaneously—dying on the sacrificial threshold. Moreover, the Upaniṣad tells us that Naciketas makes his observation of the cows after *śraddhā* enters him. Does one need *śraddhā* in order to see that the cows are old and feeble? Being *śraddhā*-inspired, Naciketas sees through the 'physical reality' into a hidden essence of reality, a vision involving a challenge to the sacrifice. Thus, Naciketas' perception differs from his father's. By virtue of *śraddhā* he diagnoses the human condition in its mortality. This is not new information, of course, but rather a new perspective, a novel realization. After his father sends him to death (Yama's abode), Naciketas' *śraddhā*-grounded voice is heard, again in the *śloka* metre:

[37] *Upaniṣads*, trans. by P. Olivelle, Oxford University Press, 1996, p. 232.

Bahūnām emi prathamo
Bahūnām emi madhyamaḥ
Kiṃ svid yamasya kartavyam
Yan mayādya kariṣyati

I walk the first of many
In the midst of others
What is death's assignment
By me fulfilled today?

Naciketas sees himself mortal, like other human beings. This may not be a trivial vision; it involves seeing the distant past and future concomitantly. Naciketas' is a sort of supra-normal perception—*siddhi*—seeing mortal humanity so clearly. It is indeed a vision similar to his perception of the 'bad-looking' cows; both visions are grounded in *śraddhā*. Naciketas' speech in the *śloka* metre, made possible by *śraddhā*-inspired perception, expresses a lonely, sudden, and subversive awakening in the course of a stressful, overwhelming situation. The young boy forcefully awakens—similarly to the Buddha—to the incurably mortal condition of man; the boy is burning with curiosity about the meaning of death. The actual condition of the healthy cows highlights the value and contents of Naciketas' spiritual perception.

But the father, untouched by *śraddhā*, does not understand. Silent, overwhelmed by the sacrifice and its odds, perhaps somewhat unfulfilled, possibly frustrated (fearing imminent failure), sick unto death as he proceeds with the sacrifice which possibly nears its end, he hears in his son's sure voice a bright echo of his inner doubt. In pain and agitation—not necessarily in anger—he cries: 'To death I give you.' Thus offering his son, he consummates the sacrifice.

Thus Naciketas does not view the actual cows presented to the priests as *dakṣiṇas*, and his unsettling, somewhat ambiguous question—'To whom will you give me?' makes radical sense. Both father and son seem to perceive failure in transcending mortality by the performance of sacrifice. However, their voices are totally different; Naciketas' voice—nourished by *śraddhā*—expresses lucid realization of the collapse of an entire culture; there is hopelessness or even sarcasm in his voice. Even if you give me as gift, what gain could be yours? How, father, can you obtain immortality by mortal means? Father, to whom will you give me? His father, sharing the sense of failure, stumbling for a solution (or answer) turns in agony to the consummation of the sacrifice. Thus, the familiar 'conflict of tradition' is acutely present in this situation. Vājaśravas is committed to

Vedic immortality; but paradoxically, in the process of sacrifice he forsakes his son, and thereby, a major expression of immortality (offspring). The father is helplessly unable to see the reality revealed to Naciketas in his *śraddhā*-trance. The cows are healthy and beautiful; what else can he do? In other words, Vājaśravas sacrifices his own son, his mode of Vedic immortality, for another type of (ambiguous) immortality. This is the way that Naciketas 'goes to the abode of Yama' (in Radhakrishnan's words). Such is the profundity of Naciketas' question; it challenges the very essence of his father's sacrifice, not the details concerning the adequacy of the cows as gifts.

In this reading, Naciketas does not wish to repair the sacrifice but, seeing through the good-looking cows into their mortality, he judges the uselessness of his father's effort to reach immortality. Thus understood, Naciketas' challenge to his father is most commensurate with the ideology of the Upaniṣads, the distinction between the narrator's voice and Naciketas' is vital in this context, and the failure to hear it has produced a series of misreadings.

The underlying theme of Naciketas' story told in the *Kaṭha Upaniṣad* is mortality. Like Maitreyī—whose life story is similarly focused on the transition from mortality to immortality—Naciketas refuses to see in property an adequate replacement for remedial knowledge. He achieves a deeper sense of inferiority which only knowledge can cure. Thus he says to Yama, the King of Death:

Since the passing days of a mortal, O Death,
Sap here the energy of all the senses;
And even a full life is but a trifle;
So keep your horses, your songs and dances!

With wealth you cannot make a man content;
Will we get to keep wealth, when we have seen you?
And we get to live only as long as you will allow!
So, this alone is the wish that I'd like to choose.[38]

Thus, Naciketas' talk with Yama re-enacts his perception of mortality which underlies his encounter with his father. He asserts to Yama a proposition equal to his criticism of his father's sacrifice: '... By fleeting things one cannot gain the perennial.'[39] Within the tension of inferiority, Naciketas experiences mortality. His question to Yama is inherently connected with

[38]*The Early Upaniṣads*, trans. by P. Olivelle, Oxford University Press, 1998, p. 381.
[39]Ibid., p. 383.

his newly awakened inferiority. He sees through his father's and the cows' inferiority, identifies it as his own, and moves on to elicit remedial knowledge from Yama. This is knowledge of the *ātman*, the one field of subjectivity. Naciketas' is a journey from mortality to immortality, from inferiority to excellence. The knowledge imparted to Naciketas echoes his sense of inferiority.

The Upanishadic stories have been under-read in accordance with the needs and identifications of those who have read them. S. Radhakrishnan, as we have seen, heavily conflates the narrative voice with Naciketas'. Doing so, he speaks for (instead of) Naciketas, for his father, and for the narrator. Radhakrishnan is thus a great scholar misled by aggressive under-reading (through conflation of textual voices); as such, his approach to the Upanishadic narrative is worthy of closer examination.

Radhakrishnan presents Naciketas as a good boy, 'disturbed by the unreality of his father's observance of the sacrifice'. Why unreality? Obviously, giving poor *dakṣiṇās* is ineffective. Radhakrishnan describes Naciketas as a naturally good boy who accepts his father's values *and* as a spiritual youth who has his own spiritual perspective:

> The author attempts to distinguish between Vājaśravasa, the protagonist of an external ceremonialism, and Naciketas, the seeker of spiritual wisdom. Vājaśravasa represents orthodox religion and is devoted to its outer forms. He performs the sacrifice and makes gifts which are unworthy. The formalism and the hypocrisy of the father hurt the son.[40]

Radhakrishnan's judgmental attitude towards Vājaśravas is conspicuous; the father is tainted by hypocrisy, external ceremony, and formalism, and is devoted to external manifestations of religiosity. The son—Naciketas—is sensitive and spiritual. Thus Radhakrishnan derives his vivid description (and perception) of a bitter, universal conflict between father and son.

However, Radhakrishnan does not allow such a conflictual situation to persist and evolve for long. He seeks to overcome his preliminary sense of hurt and challenge prevailing in the sacrificial arena, and to recompose the story in accordance with a more peaceful setting. Radhakrishnan is eager to see in Naciketas not only a spiritual figure (superior to his father), but also a very good boy. Thus he locates Naciketas within the context of education, youth and merit:

[40] *The Principal Upaniṣads*, edited by S. Radhakrishnan, George Allen & Unwin Ltd., 1953, p. 595.

Anticipating the teacher's or the parents' wishes and carrying them out is the way of the best pupils or sons; promptly attending to what is ordered is the next best; neglecting the orders is the worst form of conduct of pupils or sons. Naciketas belonged to the first type; at worst to the second; he was never negligent of his duty to his father.[41]

Radhakrishnan seems in dire need of a very good Naciketas. He asserts for Naciketas the intention to offer himself as a gift to the priests (considering the cows' poor state). This assertion derives from his conflation of Naciketas' voice with the narrator's. However, it is not impossible that a different reading haunts and bothers Radhakrishnan; perhaps Naciketas is not such a good boy, either of the first or second type of sons and pupils. And perhaps—consequently—an abysmal conflict between father and son unfolds here. Be that as it may, Radhakrishnan goes as far as to suggest that Naciketas is willing to commit suicide for the sake of his father's well-being:

The boy earnestly wishes to make himself an offering and thus purify his father's sacrifice. He does not discard the old tradition but attempts to quicken it. There can be no quickening of the spirit until the body dies.[42]

Thus, Radhakrishnan, apparently in his own independent voice, seems to suggest that death is good for the 'quickening of the spirit', in addition to assisting in a father's inadequate performance of sacrifice. A clearly Christian sensibility resonates in Radhakrishnan's voice. W. Halbfass illuminates the Christian influence one senses in Radhakrishnan's reading:

One model for the encounter of the Christian missionaries with Indian religious life and thought which gained increasing significance during the nineteenth century was the idea of fulfilment, which Nobili had already alluded to. This idea held that Indian religious concepts and convictions were not to be refuted and dismissed, but instead ought to be led beyond their own limitations to a perfection and fulfilment which the Indians themselves were incapable of seeing without being awakened to it by the Christian missionaries.[43]

However, Radhakrishnan's Naciketas is patently a self-contradictory character; a most spiritual boy, opposed to his father's mere formalism of sacrifice but also fully committed to the success of his father's endeavour:

[41]Ibid., p. 597.
[42]Ibid., p. 596.
[43]W. Halbfass, *India and Europe*, SUNY, 1988, p. 51.

Naciketas reveals here, with the enthusiasm of youth, the utter inadequacy of a formal soulless ritualism. The idea of complete surrender (*sarva-vedasaṃ dadau*) in the first verse should be properly interpreted as utter dedication or complete self-giving.[44]

And yet, in the context of this harmonious—even symbiotic—father-son relationship, there is one disturbing note: Vājaśravas informs his son of his (the father's) intention to put him (the son) to death (!) (*mṛtyave tvā dadāmīti*). The father's anger is inexplicable, unless, of course, there is no anger and the father collaborates in Radhakrishnan's programme for the quickening of Naciketas' spirit.

Radhakrishnan's reading of Naciketas' story covers the Upanishadic voice with thick veils of apologetics, conflict avoidance, and superficial good intentions. Contradictions in his presentation are obvious; if the father is poor, and gives *all* his possessions—as Radhakrishnan maintains—why is Naciketas so badly hurt by the father's hypocrisy? And if Naciketas is so spiritual, why can he not simply forgive his father for merely external elements such as the quality of the cows? Indeed, Radhakrishnan's reading is reductive in many ways, eliding textual complexity in service of hermeneutic harmony. We might ask, indeed, how Naciketas reaches the house of the King of Death? He should die first. But does he? This formal necessity—dying—is avoided in all translations of (and commentaries on) the *Kaṭha Upaniṣad*. P. Olivelle says that Naciketas' father 'gives him over to death'.[45] But how does one give another over to death? Does he kill his son? Sacrifice him? This is but one of several unattended textual gaps which emerge for the reader willing to stay with the story, a reward of untapped exegetic resources for readers who resist the seduction of seamless resolution. Consider too narrative possibilities *not* selected: Vājaśravas could have *praised* his son's *śraddhā*-inspired perception of the inadequacy of his sacrifice; he could also have appreciated his son's readiness to die for his father in an effort to repair the sacrifice. Why was he so angry then? Reading the story and its multi-vocality, the answer to this question is simple enough: the father does not share his son's perception of the cows, and consequently does not understand Naciketas' question. The cows are beautiful, healthy, and they are all he has. His response reflects his failure to understand his son's criticism. Much critical response has failed similarly, conflating

[44]*The Principal Upaniṣads*, edited by S. Radhakrishnan, George Allen & Unwin Ltd., 1953, p. 596.

[45]*Upaniṣads*, trans. by P. Olivelle, Oxford University Press, 1996, p. 411.

and thus not hearing textual voices, under-reading vocal multiplicity and thus failing to understand.

Under-reading is a strategy which serves the needs of several reader types, while doing disservice to the story itself, as we have seen: the reformer who summarizes and gets on with the job, the scientist whose ideal of objectivity necessitates objectification, and the moralist whose concept of good dictates yet another version of the story. Still, the invitation remains to return and to read differently, to resist the linear tendentious rush of reading towards resolution in favour of a more residential reading, a sustained textual cohabitation with the Upanishadic story, a reading strategy essentially experiential rather than one reductively essentialist.

FIVE

Colourless Words or Contextual Hermeneutics: the Visible and Invisible Narratives of Chāndogya 6

The Upanishadic world and experience are distant, hardly touchable. However suffused with essential otherness, Sanskrit verbalization does occur there, in the hot crucible, reflecting on and expressive of the Upanishadic reality. These verbal notes reach the ears of margin-dwellers primarily through a few key words with reference to the Upanishadic experience: brahman, knowledge, joy, *ātman*, unity (*ekatva*), the Absolute. In this distant verbalization, some aspects of the experience are somewhat subdued; wilderness, individuals, interactions, crisis, and narrative recede on the margin to inactive, unattractive background. Indeed, the given narrative is often minimal, as if inviting disregard and neglect.

The story told in *Chāndogya* 6 has become famous worldwide among Upanishadic narratives of transformation and wisdom. It contains the three words (*Tat tvam asi*) which still reverberate in the margins, each so 'colourless', as Oldenberg notes, known in India and the world as the epitome of Indian wisdom. Strangely enough, the story in which the three words are embedded is rarely attended to. The names of the heroes (Uddālaka and Śvetaketu)—as characters not philosophically relevant—are often not mentioned; the fact that Śvetaketu is his father's son is never addressed as meaningful in any way. But is this fact—Śvetaketu's filial relationship with Uddālaka—negligible, accidental?

Indeed, the story told in *Chāndogya* 6 is an illustration of a prominent much read story which is nevertheless essentially under-read. However, the transmitted story of *Chāndogya* 6 (lean, minimal) is not the full one (rich, interesting). Though one of the better narratives of uneasy interpersonal encounters leading to liberating insight among the Upanishadic tales, it is still an example of gross under-reading. However, even as an under-read story it retains much interest and is worth telling.

A father by the name of Uddālaka Āruṇi sends his son to study the Veda at a guru's house. He explains that in his Bramanical family the norm is that one becomes a real brahmin by studying the Veda. One is not a brahmin in name only. At the age of twelve, the son (Śvetaketu) goes to study the Veda. Twelve years later he returns home, proud of his knowledge and conceited. Noticing this, the father asks him: 'Did you ask for that teaching by which the unheard becomes heard, the unknown becomes known?' The son answers in the negative, and says: 'Teach me.' The father teaches him about clay and things made of clay, about copper and things made of copper, about iron and things made of iron. Clay is real, things made of clay are mere names. By knowing clay one knows all the things made of clay. The son says: 'The teachers apparently did not know this; for they would have taught me if they did! Teach me more.' And Uddālaka Āruṇi teaches of creation ('in the beginning there was one being only, one without a second ...'), of the divine breeding of multiplicity, of the constitution of man. In the course of his speeches he repeats nine times: 'That subtle being is made of self, and is everything; it is real and self, and it is you, Śvetaketu' (*sa ya eṣo 'nimaitad ātmyam idaṃ sarvaṃ tat satyaṃ sa ātmā tat tvaṃ asi śvetaketo*).

However, as suggested above, the transmitted narrative in which the most famous teaching in the Upaniṣads—*Tat tvam asi*—is embedded is not the full narrative. Composed more than 2500 years ago, the full narrative contains much which is invisible to us. Compared with margin-dwellers like ourselves, the original recipients of the Upanishadic speech-act were—naturally—connoisseurs. We, for example, may not sense the significance of Śvetaketu's age (twelve years) as he leaves his father and goes away for Vedic studies. Is this normal? Or is it rather very late for a brahmin boy? It is plausible to maintain that the full story of the communication between father and son (at the very beginning of the story) is affected by the significance of the boy's age. If twelve is very late, a brahmin child may be seen as 'fallen', a future *vrātya*, a terrible blemish and exception in the family of the greatest Upanishadic sage. The full story may tell us this but, since it is not explicit in the story, we do not perceive this in the transmitted story. Thus, we may overlook the suggestive power and insinuation implied by Uddālaka's reference to the non-existence of brahmins in name only (*brahmabandhu*) in the family. Who—in this context—is the *brahmabandhu*, and what does it mean? Can the brahmin in name only be Śvetaketu himself? Does Uddālaka charge his son with being a *brahmabandhu*, and drive him away from home against the boy's will?

Or consider the presence of Śvetaketu's teacher (or teachers), in whose home he stayed for twelve years. This presence is noted by Śvetaketu, as he says: 'These reverend sirs did not know it; for otherwise they would have told me!' How does a stay in such presence affect Śvetaketu's relationship with his father on his return home? Is it perhaps connected with the son's conceit? How is the association of fatherhood and knowledge reflected in the story? Does Śvetaketu prefer another father? Is not his 'conceit' dialogical in nature (levelled against his father)?

The transmitted story in *Chāndogya* 6 says nothing—for us—of the nature of Śvetaketu's conceit (*mahāmāna*), pride in learning (*anūcāna-māni*) and stubbornness (*stabdhatva*) on his return home—having studied all the Vedas. These traits are obviously very important in the story (the narrator's voice—as well as Uddālaka's—emphatically refers to Śvetaketu's conceit). The perception of Śvetaketu's mode of being on his return home must have been clearer to the antique narrator and his audience, but for us margin-dwellers, the problem and situation are obscure. The quest for the full story is a search for lucidity and understanding, pressing the elements of the narrative to yield as much of themselves as we can see. Though less visible, the full narrative is necessarily much richer than the transmitted one, hence the transmitted story, though more accessible, should be resisted. Conflation of the two—transmitted and full narratives—produces unnecessarily impoverished under-reading.

Two modes of under-reading have been referred to so far: neglecting the subtext and details of the story, and discarding the story as irrelevant in pursuit of the 'teaching'. These two ways of under-reading the Upanishadic story are closely related. However much read, an under-read story (its details, subtext, etc., unheeded) can hardly support and sustain abstract and complex teaching. The more famous and well known (much read) a story, the more salient its under-reading.

With this in mind, we have examined the stories of Satyakāma, Upakosala, Maitreyī, and Naciketas. The more adequate the reading of the story—the more lively, complex, and full the context—the more anchored the teaching. Thus, for example, reading Maitreyī's story as a tale of conflict unfolding between husband and wife separately seeking access to 'immortality' may be powerful enough to make a contribution to understanding the apparently abstract teaching of Yājñavalkya concerning the transcendence of duality, etc. Naciketas' *śraddhā*-inspired perception of his father's mortal (beautiful and healthy) cows sheds light on consequent developments in the story (Naciketas' preference of wisdom to pleasure, his question of the fate of man after death).

Thus my main point so far is that the Upanishadic stories are richer—*qua stories*—than is usually seen and expressed, assumed, and interpreted. However, even granting this, the question is raised: how significant is this phenomenon of under-reading the Upanishadic stories? Two points are visible on this 'continuum of relevance', a point of 'utmost relevance' and one of 'utmost irrelevance'. Utmost relevance means fully embedding the teaching within the narrative, so that *vidyā* is essentially, and perceptibly, conditioned by the narrative circumstances of its articulation. Reading thus even the most abstract teaching, such as Yājñavalkya's instruction about unity to his wife ('by what can one know what?'), is also understood as a speech-act consisting of denial of the impending separation. However, we cannot stay at this point of utmost contextualization. There is always a measure of independence in human speech. Distressed, childless, and under the pressure of his wife's poignant and powerful appeal, Yājñavalkya is still *free* to speak up, to invent and create the philosophy of Vedānta, providing mankind with the 'culminating point of the Upaniṣads'.

How does the narrative whole, the context of the fateful encounter of the frustrated wife and the spiritual husband, shape our reading of the metaphysical message of non-duality? Some of the meta-materials of Maitreyī's story infiltrate the messages; a husband is not dear for the sake of the husband but for the sake of the self. A wife is not dear for the sake of the wife but for the sake of the self. Offspring are not dear for the sake of offspring but for the sake of the self. Wealth is not dear for the sake of wealth but for the sake of the self. Husband, wife, children, wealth are part of the context of the Upanishadic revelation, and they are woven into the message. And yet Yājñavalkya , in his vision of the one *ātman*, may have transcended context altogether, leaving behind the mere circumstances of discovery, offering no additional residue of explanatory power with respect to the newly emerging message.

Yājñavalkya's metaphysics cannot—obviously—be reduced to the speech-act of a troubled husband wishing to defend himself against a disconcerted wife. Yājñavalkya *does* say something about the nature of reality and self. Thus we do not remain at the point of utmost relevance for the interpretation of the Upanishadic story and message. We also reject the point of utmost irrelevance, the decontextualized transmitted 'message'. Obviously, the Upanishadic stories have been preserved and are told in conjunction with the philosophical message (*vidyā*). And as we have seen, materials from the story do infiltrate the message. But what is the meaning of this infiltration? How does one define the boundary between metaphysics and contextual metaphysics? How relevant is the narrative?

Utmost irrelevance of the story (and particularly its 'mode') for conceptualization of Upanishadic knowledge has been the reading strategy dominant in the commentatorial and scholarly traditions. The philosophy of the Upaniṣads, the teachings of the Upanishadic sages, though taken out of context, appears understandable and lucid:

> The Upaniṣads are essentially concerned with the endeavour to describe the nature of the absolute, and the rich abundance of attempts to succeed in this end, proved by the references to conflicting theories in the Upaniṣads, is clear proof of the busy mental activity of the period. From the earliest Upaniṣad we have, the view is clear that there is a unity, that it is necessary further to grasp the nature of that unity, and that the proper names of the unity are either the Brahman or the Ātman.[1]

But reaching us thus, the messages from within the Upanishadic crucible are often deemed somewhat inferior, hardly comprehensible. Sometimes great scholars notice inner tensions of the Upanishadic voice. The vast majority of margin-dwellers are sensitive and particularly open to the dimension of 'knowledge' so prominent for the Upanishadic speakers. It is the knowledge of *ātman* or self (or Self) which is involved, knowledge infected with conceptual confusion seldom acknowledged:

> Just when philosophical interest in 'man' was about to emerge, we find Indian thought displacing it by another powerful concept, the *ātman*, which was then construed by the tradition as being *my* (or *your*) *ātman*, and by the Anglophile as being my, or your, Self—as though capitalization of the 'S' was sufficient to prevent one from looking here for what the English usage, or Western philosophical concern, understands by 'self'. One continues to do that even when the *ātman* is construed, as in Advaita Vedānta, as pure, undifferentiated, all-pervading Spirit or Consciousness (*cit*). It is still held to be my (or your) Self, if not my (or your) self—the conceptual confusion is only beginning to tighten its grip.[2]

Do margin-dwellers *understand* the Upanishadic instruction? Though some sort of joy, excitement, and sense of immortality or an aura of brahman-knowledge are lacking, a sort of understanding does seem to prevail, of course. There are some cases wherein there are obvious symptoms of understanding. Thus, for example, we note A. Schopenhauer's exclamation:

[1] A.B. Keith, *The Religion and Philosophy of the Veda and Upanishads*, Harvard University Press, 1925, rep. Motilal Banarsidass, 1989, p. 516.

[2] J.N. Mohanty, *Reason and Tradition in Indian Thought*, Oxford University Press, 1993, p. 194.

How every line is full of sure, definite and throughout harmonizing significance! How out of every page confront us deep, original, elevated thoughts, while a higher and highly sacred earnestness vibrates through the whole! ... It has become the solace of my life and will be the solace of my death.

Schopenhauer's enthusiasm betokens fiery understanding, a real event of sharing something, a taste of a 'centre of meaning' common to speaker and recipient.

Most often, however, voices on the margin are less enthusiastic than Schopenhauer's. At the same time, many suggest the spiritual relevance of the Upaniṣads. There is a note of spiritual joy in Oldenberg's appraisal of *Tat tvam asi* and the greatness of the Upanishadic sages:

Or when three briefest words, each one in itself colourless, teach one to understand the same Being in one's own person in wonderful simplicity and greatness: *Tat tvam asi*, 'That you are.' When it is seen that 'No, no' is often repeated in the old texts, or that conversation between the father and the son ends again and again with that incomparable, penetrating 'That you are, Śvetaketu,' one gets the feeling that even the authors of the Upaniṣads were indeed destined to remain alive without fading from that age onwords through millenniums.[3]

However generally appreciative of the Upanishadic sages, opinionated and original, H. Oldenberg deprives Uddālaka of any trace of compassion, viewing *Tat tvam asi* under the title of 'Egoism in the Ideal of Deliverance'.[4] Succinctly, he offers his interpretation of *Tat tvam asi* and its author's character and motive:

It is true, the sage as he appears in the Upaniṣads was little worried about applying this *Tat tvam asi* to his fellow creatures. These duties and virtues of the mundane life were for him the things of the past. The reminder 'to see all beings in his own Self and his Self in all beings' rang only momentarily in his ears. 'That you are' means for him, above all, 'that I am'.[5]

Oldenberg's boldly presumed intimate acquaintance with the interior of the Upanishadic sage's mind provides us here with a most refreshing interpretation of *Tat tvam asi*. Oldenberg argues for his charge of egoism by referring to a paraphrase of Yājñavalkya's instruction to his wife Maitreyī ('The husband is not dear because he is a husband. The husband is dear

[3]H. Oldenberg, *The Doctrines of the Upaniṣads and the Early Buddhism*, trans. by S.B. Shrotri, Motilal Banarsidass, 1991, p. 106.

[4]Ibid., p. 84.

[5]Ibid.

for love of the Self'). He concludes his narrative about the essential attribute of Upanishadic sages in the spirit of his assessment that even in Yājñavalkya's more sublime sentences (such as 'through seeing, hearing, knowing the Self ...') 'it would be gratuitous to find agreement with Kant':[6]

> It sounds as if one was afraid of becoming poor by making others rich with one's treasures of knowledge: 'The father may teach this Brahman to his oldest son or to a trusted pupil: to no one else, whosoever he may be. And even if one were to give him this earth encompassed by water and filled with treasures: he should think that this is more than that.' What a way from here to there, where Buddha said, 'Opened is the door of immortality to those who have ears.'[7]

Yet a doubt remains, as Uddālaka does seem to address his son in crisis many times—apparently—for the sake of the latter; Oldenberg's leap from *Tat tvam asi, Śvetaketo* to 'that I am' remains somewhat inexplicable.

On the other side of the of altruism–egoism continuum stands S. Radhakrishnan, who says:

> Spiritual wisdom (*vidyā*) does not abolish the world, but removes our ignorance (*avidyā*) of it. When we rise to our true being, the selfish ego falls away from us and the true integral self takes possession of us. We continue to live and act in the world, though with a different outlook. The world also continues, though it is no more alien to us. To live permanently in this new consciousness is to live in eternity.[8]

Radhakrishnan goes further in his analysis, which sees in Uddālaka's instruction to his son an endorsement of democracy. W. Halbfass sums up Radhakrishnan's orientation which culminates in his interpretation of *Tat tvam asi*:

> Radhakrishnan represented the 'idea of fulfilment' in an exemplary and especially conciliatory and impressive manner, arguing that the Vedānta is 'not *a* religion, but religion itself in its most universal and deepest significance'. He saw it as providing the framework and goal for a future synthesis of all religions and philosophies and, therewith, for the resolution of ideological and political differences and the solution of social problems. Here, the basic assumption is that Śaṅkara's doctrine concerning the absolute identity of the real in brahman must find its correspondence in a social attitude concerned with unity, equality, and reconciliation, and that it should also have fundamental effects upon the understanding of caste differences. Radhakrishnan was of the opinion that the

[6]Ibid., p. 115.

[7]Ibid., p. 84.

[8]*The Principal Upaniṣads*, edited by S. Radhakrishnan, George Allen & Unwin Ltd., 1953, p. 127.

Upanishadic formulas of unity, and especially the *Tat tvam asi* ('that art thou') characterized the 'basic principle of all democracy'; and he assures us: 'Śaṅkara's philosophy was essentially democratic.'[9]

Gazing solely at the three words, the context of their articulation fades away and disappears. The father (Uddālaka Āruṇi) and the son (Śvetaketu) in crisis are forgotten, fading away into a dense background of complete fulfilment and immortality through ultimate and Absolute knowledge—charges of egoism and democracy.

Zimmer preserves the communicative character of the *Tat tvam asi* speech-act, but apart from the style (suggesting an excited plea), he does not provide any substance to the interaction:

> *Tat tvam asi* means: 'thou art to be aware of the identity of thine inmost essence with the invisible substance of all and everything'—which represents an extreme withdrawal from the differentiated sphere of individualized appearances.[10]

Commenting on the *Taittirīya Upaniṣad*, R.C. Zaehner sees in man's consciousness of eating and being eaten a 'spiritual enlightenment and experience of immortality':[11]

> To see oneself as integrated in the world-process—not only as an 'eater of food' but as the 'food' of other living creatures—is to transcend individuality and to conquer death; for if it is true that the Being in the individual person and the Being in the sun are one (TU, 3.10[4]), then each individual man must partake in the immortality of the whole. This is a mysticism of union and communion with the totality of existence, not one of withdrawal and detachment as in the classical Yoga.[12]

Cautious and responsible, minimally dissociating himself from the Taittirīya voice, Zaehner confines his 'eater of food' and Upanishadic 'food' within distancing quotation marks, and suggests that the Upanishadic vision of union and immortality is an experience corroborated by Upanishadic reasoning ('for if it is true that the Being in the individual person and the Being in the sun are one ...')

Commenting on the same passage of the *Taittirīya*, S. Radhakrishnan speaks with assurance of the nature of Upanishadic knowledge which transcends discursive thinking and is nevertheless coherent:

[9]W. Halbfass, *Tradition and Reflection*, SUNY, p. 379.
[10]H. Zimmer, *Philosophies of India*, Meridian Books, 1957, p. 361.
[11]R.C. Zaehner, *Hinduism*, Oxford University Press, 1966, p. 51.
[12]Ibid.

A deeper principle of consciousness must emerge if the fundamental intention of nature, which has led to the development of matter, life, mind, and intellectual consciousness, is to be accomplished. The son finally arrives at the truth that spiritual freedom or delight (*ānanda*), the ecstasy of fulfilled existence is the ultimate principle. Here the search ends, not simply because the pupil's doubts are satisfied but because the pupil's doubts are stilled by the vision of Self-Evident Reality. He apprehends the Supreme Unity that lies behind all the lower forms. The Upaniṣad suggests that he leaves behind the discursive reason and contemplates the One and is lost in ecstasy. It concludes with the affirmation that absolute Reality is *satyam*, truth, *jñānam*, consciousness, *anantam*, infinity.[13]

The margin-dweller here (Radhakrishnan) momentarily conflates his own voice with that of the Upaniṣad. When he says that 'a deeper principle of consciousness must emerge if the fundamental intention of nature ... is to be accomplished', he seems to become one with the Upaniṣad through his own vocabulary and discursive power. Sometimes, however, such a great scholar, somewhat dubious about his own mystical capability, entangled in discursive thought, seeks help and refuge by resorting to poetic utterances taken from a different tradition; thus, distrusting himself, as it were—as he reaches the concept of ecstasy—S. Radhakrishnan quotes Jalaluddin Rumi:

I died a mineral and became a plant,
I died a plant and rose an animal,
I died an animal and I was a man.
Why should I fear? When was I less by dying?
Yet once more I shall die as man, to soar
With the blessed angels; but even from angelhood
I must pass on. All except God perishes.
When I have sacrificed my angel soul,
I shall become that which no mind ever conceived.
O, let me not exist! For Non-existence proclaims,
'To him we shall return.'[14]

Most often, however, understandings of the Upanishadic teaching sound less engaged and excited. However, scholars deeply interested do produce focused and illuminating descriptions of Upanishadic thought and development:

[13]*The Principal Upaniṣads*, edited by S. Radhakrishnan, George Allen & Unwin Ltd., 1953, p. 57.

[14]Ibid.

To some scholars it would appear that Vedic resemblance, governing a plethora of interlocking hierarchical registers, found its necessary end in the monistic thought represented in the early Upaniṣads. For these students of Indian history, the latter texts are continuous with, and are the teleological conclusion of, the earlier ritualistic thought. In the Upaniṣads, universal resemblance is brought to its logical terminus: universal identity. The complex system of connections between resembling phenomena, the web of *bandhus* integral to Vedic ritualism, and hierarchical distinctions are collapsed in monistic thought into the ultimate connection: the equation of self and cosmos (without the ritual intermediary) formulated as the identity and full equality of *ātman* and the *brahman*.[15]

Upanishadic otherness, however, fades away under the lucid objectivity of scholarly erudition. Margin-dwellers only seldom find leisure and inner space commensurate with reflection on the puzzling relation of the margin and the centre. Sometimes a margin-dweller wins a glimpse into the value of Upanishadic otherness, and awakens to the reality of the margin-dweller's position and the demands made of him, stooping over the Upanishadic crucible. After a somewhat awkward translation of the *Tat tvam asi Śvetaketo* paragraph (see below), M. Müller says:

> No doubt this translation sounds strange to English ears, but as the thoughts contained in the Upaniṣads are strange, it would be wrong to smooth down their strangeness by clothing them in language familiar to us, which, because it is familiar, will fail to startle us, and because it fails to startle us, will fail also to set us thinking.[16]

M. Müller, 'the doyen of nineteenth-century Indology',[17] expresses his view and values of scholarship in the qualifications he sets for true understanding of the Upaniṣads. Above all, thought, speculative thought, effort of thought, and knowledge are needed for the margin-dweller to become a real Upanishadic sage:

> To know oneself to be the Sat, to know that all that is real and eternal in us is the Sat, that all came from it and will, through knowledge, return to it, requires an independent effort of speculative thought.[18]

[15] B.K. Smith, *Reflections on Resemblance, Ritual and Religion*, Oxford University Press, 1986, p. 194.

[16] *The Upaniṣads*, trans. by Max Müller, rep. Dover Publications, 1962, p. xxxvi.

[17] B.K. Smith, *Reflections on Resemblance, Ritual and Religion*, Oxford University Press, 1986, p. 32.

[18] *The Upaniṣads*, trans. by Max Müller, rep. Dover Publications, 1962, p. xxxvi.

In the introduction to his translation of the Upaniṣads Müller suggests that he himself has 'realized' the thoughts of the Upanishadic sages, having known the ancient truths 'from within' rather than 'from without'.

We must realise, as well as we can, the thoughts of the ancient *Rishis*, before we can hope to translate them. It is not enough simply to read the half-religious, half-philosophical utterances which we find in the Sacred Books of the East, and to say that they are strange, or obscure, or mystic. Plato is strange, till we know him; Berkeley is mystic, till for a time we have identified ourselves with him. So it is with these ancient sages, who have become the founders of the great religions of antiquity. They can never be judged from without, they must be judged from within.[19]

But how does one think the Upanishadic thought from within? Should we—margin-dwellers—*become* Upanishadic rishis? Müller suggests that, indeed, 'we must for a time' become such seers of antiquity. This is true, Müller says, especially for people like himself (who venture to translate ancient scriptures of the East).

We need not become Brahmans or Buddhists or Taosze altogether, but we must for a time, if we wish to understand, and still more, if we are bold enough to undertake to translate their doctrines.[20]

Müller's is one of the more sincere and valuable reflections on the condition of marginality in relation to the Upanishadic thought (and experience). Müller's self-understanding as a bold margin-dweller alternately becoming a Brahman (or a Buddhist) and then returning to residency on the margin captures an essential truth about scholarly existence. He seemingly addresses other margin-dwellers who are not bold enough to become—even if only for a time—Brahmans, Buddhists, etc.

Whoever shrinks from that effort, will see hardly anything in these sacred books or their translations but matter to wonder at or to laugh at; possibly something to make him thankful that he is not as other men.[21]

Müller's suggestion that we become Brahmans sometime so that we may understand Upanishadic thought from within is an admonition to other—apparently complacent and judgmental—margin-dwellers who shrink from the effort at Brahmanic understanding. Müller's criticism of his fellow

[19]Ibid., pp. xxxvi–vii.
[20]Ibid., p. xxxvii.
[21]Ibid.

margin-dwellers may also be a kind of self-criticism of one who has not become a Brahman, remaining on the margin. Müller then apparently injects his own version of identity or unity (of mankind, seemingly), which nourishes a sense of confidence in his own understanding of sacred texts such as the Upaniṣads:

> But to the patient reader these same books will, in spite of many drawbacks, open a new view of the history of the human race, of that one race to which we all belong, with all the fibres of our flesh, with all the fears and hopes of our soul.[22]

Latent love for the ancient East vibrates in Müller's vision of the unity of mankind; properly reading the Upaniṣads may mean understanding this unity, sharing in a vision of a single humanity. Müller's rupture is not exactly an Upanishadic one (since he apparently commits himself to the mere unity of mankind), and yet he is driven to a bold formulation of this unity:

> We cannot separate ourselves from those who believed in these sacred books. There is no specific difference between ourselves and the Brahmans, the Buddhists, the Zoroastrians, or the Taosze.

So it seems—margin-dwellers may wonder—that one does not *have* to become a Brahman, after all, for one's identity with the Brahmans is already given. There are minor differences, of course, as negligible as can be:

> Our powers of perceiving, of reasoning, and of believing may be more highly developed, but we cannot claim the possession of any verifying power or of any power of belief which they did not possess as well. Shall we say then that they were forsaken of God, while we are His chosen people? God forbid![23]

Müller extricates thought from the context of its occurrence. The world of thought is one, he suggests. While we may not in all probability really become Brahmans, we may become Thinking Brahmans, Brahmans in thought and ideas. Thus Müller earns his meticulous rendering of the most famous teaching in the Upanishadic literature (*sa ya eṣo 'ṇimaitad ātmyam idaṃ sarvaṃ tat satyaṃ sa ātmā tat tvam asi svetaketo*). The cost of extricating thought from context can be seen in his paraphrasing of Uddālaka's teaching; it is lifeless, self-conscious, unnecessarily long and repetitive, abstract, and with no trace of true excitement:

> That which is the subtle essence (the Sat, the root of everything), in it all that exists has its self, or more literally, its self-hood. It is the True (not the Truth in the

[22]Ibid.
[23]Ibid.

abstract, but that which truly and really exists). It is the Self, i.e. the Sat is what is called the Self of everything. Lastly, he sums up, and tells Śvetaketu that, not only the whole world, but he too himself is that Self, that Satya, that Sat.[24]

Müller's exposition of the *Tat tvam asi* discourse—though strange—'fails to startle us' and is thus somewhat disappointing. Given his self-understanding as an occasional Brahman and his personal version of unity, one would expect a more experiential, self-assured and powerful rendering of this discourse. Avoiding the story and characters, wandering in a universe of thought and capital letters, Müller is lost at this point, unable to touch Śvetaketu, the exiled, distressed son who returns home, the aggressive yet compassionate father, the crisis and its resolution expressed in the famous text.

As Müller says, the power of knowledge (of 'oneself to be the Sat, to know that all that is real and eternal in us is the Sat, that all came from it and will, through knowledge, return to it') is a major theme in the Upanishadic literature. The association of knowledge and *ātman* is, indeed, central in the Upaniṣads. It is the Upanishadic drama *par excellence*, its moment of heroism, enlightenment, and conversion. Knowledge of *ātman* is precious, powerful, redeeming, and is kept secret. People who devote themselves in the wilderness in the quest for knowledge go the way of the gods (*devayāna*) after death.[25] King Death, Yama, is unwilling to tell Naciketas the fate of man after death;[26] King Jaivali Pravāhaṇa is similarly reluctant to convey knowledge to Uddālaka Āruṇi, a great brahmin portrayed as a lover of knowledge, hating pretension to knowledge and ready to receive instruction from a king (*kṣatriya*).[27]

The power and significance of knowledge (*vidyā*, *jñāna*, *vijñāna*) is the dominant theme of *all*—without exception—stories told in the Upaniṣads. But what is knowledge in the Upaniṣads? Having read the Upaniṣads, are we knowers of Upanishadic knowledge? I think this question is a serious one. While the answer is somewhat obvious (we do not become knowers of Upanishadic knowledge by reading the Upaniṣads), the question about the nature of *vidyā* becomes more accessible as we concede ignorance. Closest is the location of the self-conscious, reflectively ignorant dweller on the margin; for he forsakes pretensions to knowledge, a pretension shared by anyone who describes the *contents* of the knowledge contained in the Upaniṣads.

[24]Ibid., p. xxxvi.
[25]See ChU 5.3. and BU 6.2.
[26]See KU 2.
[27]ChU 5.3.7.

Most margin-dwellers, though they do not confess to having become Brahmans (even for an hour), speak in a voice of scholarship and thought. This voice has its own ring and melody:

By the time of the Upaniṣads the word *brahman*, though it retained its original meaning of 'sacred utterance' in the specialized senses of both the Veda and the Brahman class which was the guardian of the Veda, ordinarily meant the Absolute, that is, what remains unchanged in a world of change. This, however, was not enough for the Upanishadic sages, and they tried to define more accurately what this something was. For the *Taittirīya Upaniṣad* the basic element was food, for the *Kaushītaki* breath. In more modern terminology it was matter for the former, spirit for the latter.[28]

A standard paraphrase and explanation of Upanishadic knowledge—in the form of explaining *Tat tvam asi*—is M. Hiriyanna's, referring to the 'happy identification' of *ātman* and brahman, after these two were conceptualized independently before the Upanishadic era:

Thus Brahman means the eternal principle as realized in the world as a whole; and ātman, the inmost essence of one's own self. These two conceptions—Brahman and ātman—are of great importance and occur not only independently in the literature of this period, but are sometimes correlated with each other. ... The two conceptions are also sometimes identified; and it is this happy identification of them which constitutes the essential teaching of the Upanishads. It is represented by the well-known sayings 'That thou art' (*Tat tvam asi*) and 'I am Brahman' (*Ahaṃ brahmāsmi*). They mean that the principle underlying the world as a whole, and that which forms the essence of man, are ultimately the same.[29]

Hiriyanna has forgotten the protagonist speaker (Uddālaka) and his audience (Śvetaketu). He even strips *Tat tvam asi* of its immediate philosophical environment, which consists of the repetitive reference to the subtle (*aṇima*) essence underlying the entire world. Śvetaketu, the son driven away from home and returning there imbued with a sense of superiority, challenging his biological father (Uddālaka) and defeated in turn (by his father) in a Vedic contest, is totally forgotten in standard expositions of the most famous occurrence of the Upanishadic philosophy and teachings. Most often, scholars mention the names of the spiritual heroes involved, but not their dramatic situation, conflict, character, and story; thus, for example:

[28]R.C. Zaehner, *Hinduism*, Oxford University Press, 1966, p. 51.

[29]M. Hiriyanna, *The Essentials of Indian Philosophy*, George Allen & Unwin, 1985, p. 21.

The *Chāndogyopaniṣad*, like the *Bṛhadāraṇyaka*, contains a wide variety of teachings. One of the most important of these, and the one most often referred to in later times, is the instruction given by the Brahman householder Uddālaka to his son Śvetaketu. The central theme of the instruction is the difference between the underlying reality and the modifications in that reality that give rise to manifold names and forms. Uddālaka explains to his son that the truth (*satya*) of anything made of clay is clay, copper is the truth of what is made of copper, and iron the truth of all the things made of iron. So is it also, he says, with the universe. The universe in the beginning was Being or the Existent (*sat*) alone, One without a second; from that the plurality of existent entities was born by successive modifications of the One. The One created or emitted heat, heat created water, and water, food. The highest power, Being itself, entered by means of the living self (*ātman*) into the three created potencies (heat, water, and food), causing them to combine in various threefold ways and become differentiated in name and form. In this way, though it was in truth the one sole origin and highest self, the Existent became also the self of all individual created beings.[30]

This reduction to contents invites comparison with other contents found in different philosophical traditions. Reduction to contents along with a mood of comparison and a courageous, independent note of judgement is found in P. Deussen's account:

This is the oldest passage in which the unreality of the manifold world is expressed. Not long after this, Parmenides in Greece attained to the same knowledge and uttered it almost in the same way: 'That is why everything which trustful men have accepted as truth is all what arises and disappears, a mere name.' Spinoza gives expression to the same knowledge when he explains all individuals as modes ('modi') of a divine substance ('substantia'). All the three—Chāndogya, Parmenides, Spinoza (and thus all philosophers before Kant) commit the error to transfer the empirical forms of ideas ...[31]

The reduction of Upanishadic knowledge to 'contents' has been supplemented by another reduction, the most aggressive textual essentialism imaginable: the representation of one or two sentences as the epitome of Upanishadic wisdom. We have in mind, of course, the paradigmatic *mahāvākya*, *Tat tvam asi*. Here is the standard approach, expressed by a leading authority:

[30]T.J. Hopkins, *The Hindu Religious Tradition*, Dickenson Publishing Company, 1971, p. 43.

[31]P. Deussen, *Sixty Upaniṣads of the Veda*, trans. by V.M. Bedekar and G.B. Palsule, Motilal Banarsidass, 1991, p. 156.

Really striking is the directness with which here the whole essence of the mysterious highest Being, the entire fullness of the godhead or divinity has been recognized in the discourse; the words *Tat tvam asi* are rightly regarded as valid embodying the sum-total of all Upanishadic teaching.'[32]

Though beautifully economical and also—possibly—promising fast awakening (just concentrate on *Tat tvam asi*), this approach betokens disregard for the nature of Upanishadic knowledge which may be incomparable—in principle—to other types of knowledge. For indeed, apart from happy identifications of *ātman* and brahman, recognitions of the entire fullness of the godhead or divinity, the Upaniṣads profusely testify to complex problem situations revealed and expressed in highly suggestive and powerful stories. Such stories point to the mode of reality and the nature of Upanishadic knowledge, and constitute an opportunity for margin-dwellers to 'become Brahmans for a time' (if such temporary change of heart—envisaged and desired by M. Müller—is possible). Yet it has been precisely those stories that have repeatedly been left behind in cursory readings racing through them toward perceived philosophical truths.

The Upanishadic story is indispensable to learning about the nature of Upanishadic knowledge. For obviously, the knowledge of 'unity' expounded in the Upaniṣads is not like other 'knowledge'. It is mysteriously—and abysmally—different from many kinds of information we possess. One is in need of inner space commensurate with this knowledge, inner space dearly acquired and maintained.

And who could tell us something about this better than the Upanishadic narrative? For the nature of knowledge is only partially indicated by the contents of knowledge. Indeed, under-reading the Upanishadic narratives deprives the Upanishadic experience (and nature of knowledge) of its essential otherness. Many, however, are the great minds who attempt to define an essence of knowledge, an essence found in all types of knowledge. Such for example is J.N. Mohanty's penetrating and useful explication:

To be the subject of knowledge requires transcending one's personal interests and prejudices, and to attain universality, such that knowledge is, in principle, valid for everyone. Thus the epistemological subject is disinterested and also universal, for otherwise knowledge could not be objective and could not be valid for everyone. It is not being denied that there is a perfectly legitimate sense in which it is a person who knows. But, if a person's claim to know is to be sustained, then his

[32]Ibid., p. 159.

being the person that he is should be irrelevant to his knowing what he knows. For what he knows should be an objective truth, which could be known by any other person as well. His knowing it requires that *qua* knower he is disinterested, even if he might have been led to his knowledge by interests and his may give rise to further interests.[33]

Mohanty's definition highlights important aspects in the phenomenology of knowledge as this concept of knowledge is currently used. Some of these aspects are significantly enlightening about—and even particularly befitting—the nature of Upanishadic *jñāna*. However, there is also *otherness* embedded in the Upanishadic *vidyā* or *jñāna* as well as in the nature and significance of Upanishadic experience; such otherness is in danger of being totally overlooked. Mohanty's striking distinction between 'person' and 'subject' deprives the storytelling which deals primarily with mere 'person', of its power and meaning, providing a coherent justification for the neglect of the Upanishadic story and its under-reading. In his 'Philosophy as Reflection on Experience', he says:

Philosophy, in my view, consists of reflection on man's experience in relation to himself, to others and to the world. However, as a philosopher I find myself in a curiously ambivalent position. Being steeped in history, my reflection presupposes and proceeds on the foundation of the sedimented thoughts of other philosophers who either have preceded me in time or, through my contemporaries, have been able to exercise some significant influence on me. As a consequence, one may either say that my access to my own experience is mediated by the history of thought or that my experience itself is constituted by that history.[34]

In this view, the Upanishadic story is ideologically and experientially redundant. Mohanty's dissociation of the subject from the person and his self-understanding as a philosopher living solely in textual space prevent him from seeing the multidimensional otherness of the tradition of thought and experience he comments upon. The reality and otherness of the Upanishadic universe is transmitted, above all, by the storytelling and narrative of the Upaniṣads.

The Upanishadic stories are essentially dialogical, knowledge being transmitted in a context of dialogue, an interpersonal environment. It is often a serious and charged situation; a deadly contest among great sages,[35]

[33]J.N. Mohanty, *Reason and Tradition in Indian Thought*, Oxford University Press, 1992, pp. 195–6.

[34]J.N. Mohanty, *Essays in Indian Philosophy*, Oxford University Press, 1993, p. 1.

[35]See BU 3.

a crisis in the family,[36] etc. Moreover, the relationships wherein knowledge is transmitted often involve conflict, which is, in a sense, heightened contact with duality. In terms of the narrative, the drive to unity is a resolution of conflict. The transcendence of duality, conceptualized on the margin as a purely theoretical matter, is presented in correlation with the storytelling of conflict and its resolution. The Upanishadic stories are full, whole, and complex, and, as such, conflict underlies many of them. Thus 'knowledge' or the transmission of knowledge is often, in the narrative, resolution of conflict.

Viewing the stories, we may witness conflict as the expression of the pain of duality. Maitreyī and Yājñavalkya, Naciketas and his father, Naciketas and Yama, Satyakāma and his mother and his teacher, Upakosala and his guru, Uddālaka and his son Śvetaketu, Jānaśruti and Raikva, are all pairs in conflict, moving to obtain 'knowledge'. This is the fundamental, pre-liberation dual reality unfolded in the Upanishadic narrative. The Upanishadic story tells of the resolution (or solution) of the conflict, the teaching of unity being the culminating point of the story. Thus the great moment in the sixth chapter of the *Chāndogya Upaniṣad* is *both* the revelation of ultimate truth (*Tat tvam asi*) *and* the resolution of conflict between Uddālaka and his son Śvetaketu. These two levels in the Upanishadic conceptualization of transcendence (of duality) are inseparable, hence the translation and interpretation of Uddālaka's 'great saying' (*Tat tvam asi*) should also be compatible with the *two* lines of exposition articulated in *Chāndogya* 6 (and essentially in all the cases of transmission of Upanishadic knowledge). Such a translation would thus be: 'You are me (and I am you), Śvetaketu.' The abysmal conflict between father and son is resolved as Uddālaka shares his being with Śvetaketu, transmitting (or trying to transmit) the truth which makes them one. This sharing, typically Upanishadic, powerfully illuminates the nature and kind of Upanishadic 'knowledge' grounded in conflict and pain, different from the (context-free) metaphysics received on the margin.

We turn now to exercise a contextual metaphysics, the interpretation of the Upanishadic message reflecting on the relevance of the story to metaphysics. As discussed above, the story of Uddālaka Āruṇi and his son Śvetaketu contains the most famous 'teaching' of the Upaniṣads—*Tat tvam asi śvetaketo*. As noted, a most aggressive decontextualization has been applied to this piece of knowledge, the essence of Upanishadic knowledge. In the rich tradition of Advaita Vedānta, the most prestigious among Indian

[36] Such as in the case of Maitreyī, Satyakāma, etc.

schools of thought and soteriology, the two sentences—*Tat tvam asi* and *ahaṁ brahmāsmi*—have been considered the exclusive 'great sayings' (*mahāvakyā*) of the Upaniṣads. The method of liberation has been closely associated with a correct understanding of these two *mahāvakyās* and in particular *Tat tvam asi*. Consequently, the entire destiny and future of mankind has been perceived to depend on the comprehension of *Tat tvam asi*. Thus came into being the most intense gaze possible focused on a few 'colourless'[37] yet miraculous, omnipotent words, taken completely out of context:

> All of the effort of what I call Advaita Vedānta hermeneutics is directed towards the proper understanding of these *mahā-vākyas* with a purely soteriological motive. Finding out what they mean, which is equivalent to directly experiencing their meaning, produces liberation.[38]

Most prominent and formative in this tradition of context avoidance with respect to a great saying such as *Tat tvam asi* is Śaṅkara. While Śaṅkara does not use the concept of *mahāvākya*, the presence of the two paradigmatic sentences (and in paricular *Tat tvam asi*) in his mind is most conspicuous in all of his works.[39] A steady gaze at the three words to realize their proper meaning is proposed in Śaṅkara's work:

> In BSBh 4.1.2, in the context of discussing the significance of the repetition of the statement 'That thou art' in ChU 6, Śaṅkara states that the sentence expresses an identity between the sense or content of the word 'thou' (*tvam*) and the sense or content of the word 'that' (*tat*). Further, the sense of the word *tat* is given as that Brahman which is the cause of the origin of the universe ... while the sense of the word *tvam* is the inward Self (*pratyagātman*). Śaṅkara goes on to say that, for those persons for whom the proper meaning of these two words is somehow obstructed, the repetition of the statement assisted by reasoning (*yukti*) will lead to a proper discarding (*avadhāna* or *apoha*) of the false meanings of the words *tat* and *tvam*.[40]

Obviously, a major imbalance between the story and the teaching is implied by the very concept of *mahāvākya*. The three words become the gem of Indian wisdom, detached from their immediate textual environment (the story of Uddālaka and Śvetaketu). As suggested above, the major thrust of

[37]Oldenberg's predication of the three words (taken separately). See earlier in this chapter.

[38]I. Kocmarek, *Language and Release*, Motilal Banarsidass, 1985, p. 15.

[39]See, for example, the references to *Tat tvam asi* in his *Brahmasūtrabhāṣya*.

[40]Ibid., p. 22.

the accounts given of the Upaniṣads is philosophical and de-contextualized. Thus, for example, the following account depicts Upanishadic knowledge in terms of 'theological empiricism' and 'impersonalist monism':

Uddālaka Āruṇi, who along with Yājñavalkya can be regarded as one of the earliest Hindu theologians, in dialogue with his son Śvetaketu, illustrates how *brahman* is the essence, the smallest particle of the cosmos. In an early example of theological empiricism, he splits a fruit and then the fruit's seed to show how *brahman* cannot be seen. Similarly, as salt placed in water by Śvetaketu completely dissolves and cannot be seen, though it can be tasted, so *brahman* is the essence of all things, which cannot be seen but can be experienced. This essence is the self, and the passage explicating this concludes with the famous lines: 'That which is minute, the totality is the self. That is truth. That is the self. That you are, Śvetaketu.' This impersonalist monism ...[41]

Other good accounts of Upanishadic wisdom vaguely suggest practical dimensions of knowledge, yet focus on 'knowledge' without reflection on its nature:

The teachings of Yājñavalkya and Uddālaka are the best developed statements of the early Upaniṣads. Together they exerted an enormous influence on the later tradition, not least because they represent a basically pragmatic concern. Their attempts to locate and define the self were not exercises in abstract thought or philosophy for its own sake. Their goal was salvation, defined as the release of the self from its continued bondage to rebirth.[42]

However, the emphasis is invariably on the *contents* of knowledge rather than on its nature:

They are agreed that the ultimately real, the One without a second, is Brahman, and that Brahman is present in the phenomenal world as the self of man. They are further agreed that the knowledge that brings release is knowledge of the unchanging Brahman, and that this knowledge is obscured by the condition of the embodied self.[43]

Let us now examine in more detail that most well-known episode in the philosophical tradition of the Vedānta, the story of Uddālaka Āruṇi and his son Śvetaketu as told in the sixth chapter of the *Chāndogya Upaniṣad*. Uddālaka is Śvetaketu's father, and he says to his son: Śvetaketu, go and

[41]G. Flood, *An Introduction to Hinduism*, Cambridge University Press, 1998, p. 85.
[42]T.J. Hopkins, *The Hindu Religious Tradition*, Dickenson Publishing Company, 1971, p. 45.
[43]Ibid.

stay in *brahmacarya*, for in our family there is no one who claims to be a brahmin by nature (birth) as it were (*brahmabandhuriva*), without studying (*śvetaketo, vasa brahmacaryaṃ na vai somyāsmatkulīno 'nanūcya brahmabandhuriva bhavatīti*). Or: In our family one does not become a true brahmin by virtue of his family; one becomes a brahmin by study!

A complex conflict underlies Uddālaka's opening statement. His explanation, indeed, is a somewhat lengthy one; he seems to be trying to overcome his son's resistance. For why, indeed, should Uddālaka explain to Śvetaketu the need to go and study? The *Chāndogya* is explicit on Śvetaketu's age and term of study; he is twelve and goes away as a *brahmacārin* for twelve years.[44] For a qualified recipient of the *Chāndogya*, it is time for Śvetaketu to leave home and study the Veda; why the lengthy explanation? Referring to *Āpastambasūtra* 1.1.18, M. Müller focuses on Śvetaketu being twelve: 'This was rather late, for the son of a Brahman might have begun his studies when he was seven years old.'[45] Müller adds that the age of twelve 'was considered the right time for mastering one of the Vedas'.[46] Müller's note helps our thinking about the possible circumstances evoked and embedded in the Upanishadic voice, of Uddālaka's mood, perception, and assessment of his son's condition. Śvetaketu's age (twelve) at the time of his being sent away to a guru's home suggests the event of *upanayana*, the most important *saṃskāra* of Brahmanical society, the initiation into the study of the Veda.

> At the time of initiation (*upanayana*) the adolescent boy, dressed in an antelope skin, receives from his master or *guru* the words of the celebrated verse of the *Sāvitrī*, along with the sacred thread and the staff. From that moment, which is considered as if it were his spiritual birth, he is 'twice-born' (*dvija*) and responsible for his acts.[47]

According to Manu, 'the ceremony of initiation should be performed in the eighth year after (the conception of) the embryo of a priest; in the eleventh year after (the conception of) the embryo of a king ... for a priest who desires the splendour of the Veda it is to be done in the fifth year.'[48] Thus, in our choice of the full narrative most compatible with the transmitted one, we must address the invisible yet integral part of the

[44]See ChU 6.1.2.
[45]*The Upaniṣads*, trans. by Max Müller, rep. Dover Publications, 1962, p. 92.
[46]Ibid.
[47]R. Lingat, *The Classical Law of India*, University of California Press, 1973, p. 11.
[48]*The Laws of Manu*, trans. by W. Doniger and B.K. Smith, Penguin Books, 1991, p. 21.

real story: the initiation of Śvetaketu into the study of the Veda and the significance of its failure or postponemet. For indeed, the failure to undergo *upanayana* is fateful for many generations:

Initiation is obligatory for all males of the free castes. He who abstains from it falls, he and his descendants, into the status of the 'fallen', the excommunicated. In principle initiation should be followed by a more or less long novitiate, devoted to the acquisition of the Veda by the initiate.[49]

Upanayana is the most important rite of passage for antique Vedic society, 'more or less a compulsory institution'[50] in Vedic India. The connection of such a central event in the life cycle of brahmins in Vedic India with Uddālaka's charging to Śvetaketu of being a brahmin in name only (*brahmabandhu*) shows the depth of the crisis implied and expressed in Uddālaka's initial statement.

In his description of *upanayana* and its meaning, B.K. Smith refers to the connection between *upanayana* and the concept of *brahmabandhu*, a brahmin in name only:

Although the rituals of healing and construction begin with the act of intercourse and conception, up until the time of initiation into Veda study and sacrificial ritualism (the *saṃskāra* known as the *upanayana*), the boy is considered incomplete and not yet responsible. A Brahmin, for example, who has not undergone initiation is called a 'Brahmin by birth only' or a 'Brahmin by relation only'.[51] They are in a state of nature, and may therefore 'Act naturally', but they are also prohibited from the cultural act par excellence of sacrifice: 'Before initiation [a boy] may follow his inclination in action, speech, and eating [but he shall] not partake of sacrificial offerings.' Such natural humans do not deserve the designation Brahmin at all; they are likened to wooden elephants or leather antelopes which 'have nothing but the name' and are 'equal to *śūdras*'—for the members of this latter class have not been 'reborn' into Aryan society through the initiation and therefore are known as the 'once born'.[52]

Thus, Śvetaketu may be not only 'unlearned in the Veda' but also ontologically inferior, unregenerate, once-born, a *śūdra* in his father's eyes. Viewed thus, Śvetaketu is one of the 'unregenerates and, therefore, in the

[49]R. Lingat, *The Classical Law of India*, University of California Press, 1973, p. 11.

[50]J. Gonda, *Change and Continuity*, Mouton & Co., 1965, p. 391. See B.K. Smith, *Reflections on Resemblance, Ritual, and Religion*, Oxford University Press, p. 94.

[51]B.K. Smith refers here to Śvetaketu's story in ChU 6.1.1. *Reflections on Resemblance, Ritual and Religion*, Oxford University Press, 1989, p. 87.

[52]Ibid., p. 87.

eyes of the ritualists, degenerates [who] are all regarded as less than really human, petrified in a state of non-being'.[53]

Thus heeding the meaning and significance of *upanayana* as a rite of passage, a second birth, we may better assess the emotions and meanings involved in the uninitiated, twelve-year-old Śvetaketu's 'absence from school'. Uddālaka's reference to *brahmabandhu* (a brahmin by birth) who has not studied (*ananūcya*) the Veda are significantly addressed to Śvetaketu; *he* is the one charged by his own father with being a *brahmabandhu*, potentially a *śūdra*. This is a remarkably aggressive move on Uddālaka's part, and his explanation is thus obviously not just an explanation but an accusation as well. In short, Uddālaka drives his son away from home. Conflict and resentment colour the relationship of father and son from the very outset.

Great observers have been on the verge of closely reading this story of the son's exile from his father's home. Thus a question pops up in Śaṅkara's mind: Why does not Uddālaka *himself* teach his son? The great Śaṅkara answers: we may infer that the father, though qualified (*guṇavān*) could not instruct his son because of the *father's absence from home* (*tasyātaḥ pravāso 'numīyate pituḥ yena svayaṃ guṇavān san putraṃ nopaneṣyati*). However, focusing on the technicality of being away from home, Śaṅkara's suggestion diminishes the story to triviality. There is no question of conflict or meaningful encounter or mutual criticism between father and son. Śaṅkara reduces the potential of the narrative to facilitate understanding of the message. And yet he does sense the big event lurking behind the terse narrative; Uddālaka drives his son away from home.

After twelve years of *brahmacarya*, the son returns home conceited. He thinks he is learned (*anūcāna-māni*), apparently responding to his father's accusation twelve years earlier (of his being the *brahmabandhu* who has not studied the Veda (*ananūcya*)). Śvetaketu's pride is dialogical in this context, as he apparently disproves his father's verdict of him as a fallen brahmin, a disgrace to his family. In the background one senses the presence of Śvetaketu's teacher who initiated and instructed him, and who has replaced Uddālaka as Śvetaketu's father. While such a possibility seems far-fetched, it is recognized in Hindu tradition (discussed later).

Reference to Śvetaketu's conceit, though very salient in the narrative, has been grossly under-read. Translators and scholars tend to remain satisfied with conceit as a self-explanatory fact; the boy is conceited since he thinks

[53]Ibid.

he knows so much. Śvetaketu's pride is contrasted with his father's humility. 'Śvetaketu is often depicted as a haughty young man contrasting sharply with the humility of his father.'[54] However, in the unfolding story of the conflict between father and son the latter's conceit is a turning point, the apex of the crisis. The father moves ahead and breaks the network in which his son's pride is anchored. But why is this so badly needed, from Uddālaka's point of view?

Close reading of the story at this point is essential, yet difficult. We must assume that the Upanishadic emphasis laid on 'conceit of learning' (*anūcāna-māna*) is important. In the terse and condensed storytelling of *Chāndogya* 6, there are six main facts: Śvetaketu's delinquent behaviour; his exile from home; his return, conceited; the crisis enacted by his father through the pride-breaking question (concerning the transformative teaching by which what is unheard becomes heard); the son's application for knowledge; Uddālaka's sharing 'knowledge' (teaching) with his son (including *Tat tvam asi*). Śvetaketu's return home conceited is thus one of the main facts of the story, and part of the unfolding relationship and crisis with his father.

Śvetaketu's disposition to replace his father, to choose his Vedic teacher as father, is corroborated by his suggested superiority over his biological father (Uddālaka). Thus the young man's conceit is his choice of another father together with a sense of superiority rooted in the quality of his guru. This is the background of his saying: these sages did not apparently know it, for if they did they would have told me! With these words, Śvetaketu starts on his way home, to his father Uddālaka. His conceit even implies that he may have considered his father his son; thus, according to *Manusmṛti* 2.150:

> The priest who brings about the Vedic birth of an older person and who teaches him his own duties becomes his father, according to law, even if he is himself a child.[55]

And Manu sums up: 'For an ignorant man is really a child and the one who gives him the Vedic verses is his father; people call an ignorant man, "Child", and a person who gives Vedic verses, "Father"'.[56]

And thus, Śvetaketu's conceit of learning is rooted in and nourished

[54] *Upaniṣads*, trans. by P. Olivelle, Oxford University Press, 1996, p. 414.

[55] *The Laws of Manu*, trans. W. Doniger and B.K. Smith, Penguin Books, 1991, p. 32.

[56] Ibid., p. 33.

by his education at the guru's home; the association of learning (sharing knowledge) and fatherhood suggests that Śvetaketu returns home to repudiate his father's fatherhood; this is the fuller real narrative behind the conceit reported in the given story. The teaching unfolding in the sixth chapter of the *Chāndogya* is thus a continuous assertion of fatherhood, culminating in *Tat tvam asi.* In this context, Uddālaka enters into a deadly Vedic contest with his son, and—though defeating rather than 'killing' Śvetaketu, he shares with him happiness and existence. You are not what you think, but you are that, what I am already.

Uddālaka starts his assumption of renewed fatherhood by openly challenging his son: did you ask for that transformative instruction (*ādeśa*) by which what is unheard becomes heard, the unthought of becomes thought, the unknown becomes known? (*uta tam ādeśam aprakṣyo yena aśrutaṃ śrutaṃ bhavati*, *amataṃ matam*, *avijñātaṃ vijñātamiti*) (6.1.3). It is an unsettling, baffling, paradoxical, even aggressive question, most difficult to answer.

The narrative strongly suggests that the teaching (*ādeśa*) by which *aśrutaṃ* becomes *śrutaṃ*, etc., is the *Tat tvam asi* teaching. All the commentators and scholars agree that the teaching of brahman is intended here. The promise of knowing everything by knowing one thing suits the spirit of the Upaniṣads and is expressed elsewhere. The *Muṇḍaka Upaniṣad* 1.1.3 and the *Bṛhadāraṇyaka Upaniṣad* 2.4.5 refer to the possibility of knowing the 'whole world' by knowing one thing. In the dialogue between Maitreyī and Yājñavalkya, the sage ends his memorable speech on the husband not dear by virtue of himself (but by virtue of the self) by addressing his wife: 'Maitreyī, my dear, through a sight of the self, by hearing of it, by thinking of it and understanding the self, everything is known' (*maitreyī ātmano vā are darśanena śravaṇenāmatya vijñānenedaṃ sarvaṃ viditaṃ*). In the *Muṇḍaka*, Śaunaka, the great householder (*mahāśāla*) approaches Aṅgiras and asks him: 'Knowing what, venerable sage, all this becomes known?' (*kasmin nu bhagavo vijñāte sarvam idaṃ vijñātaṃ bhavatīti*). Aṅgiras' answer is pertinent to the context of Uddālaka's question to Śvetaketu; Aṅgiras says that there are two kinds of knowledge to be known (*dve vidye veditavye*), the higher (*parā*) and the lower (*aparā*). The higher is that by which one realizes the 'imperishable' (*akṣara*) (*parā yayā tad akṣaram adhigamyate*). Lower knowledge consists of the four Vedas and the sciences of phonetics, ritual, grammar, etymology, metrics, and astrology (*tatrāparā ṛg-vedo yajur-vedaḥ sāma-vedo 'tharva-vedaḥ śikṣā kalpo vyākaraṇaṃ niruktaṃ chando jyotiṣamiti*).

Among other things, Uddālaka's baffling question to his son is an expression of the changes of culture and values established, reflected on and expressed in the early Upaniṣads. It is also, of course, a turning point in the story of *Chāndogya* 6, as Uddālaka starts making his move to reclaim his son, breaking as it were Śvetaketu's identification and affiliation with his surrogate father (his guru). Uddālaka's question is important in the narrative as well as with reference to spiritual and philosophical 'contents'.

M. Müller discusses this question in his introduction to the second volume of his translation of the Upaniṣads. In this introduction he refers generally to the difficult task of translating the Upaniṣads:

> Whatever other scholars may think of the difficulty of translating the Upanishads, I can only repeat what I have said before, that I know of few Sanskrit texts presenting more formidable problems to the translator than these philosophical treatises.[57]

Müller translates Uddālaka's question to his son thus: 'Have you ever asked for that instruction by which we hear what cannot be heard, by which we perceive what cannot be perceived, by which we know what cannot be known?'[58] Müller compares his translation with N. Goreh's,[59] R. Mitra's,[60] Gough's,[61] and Ballantyne's.[62] The crux of Müller's translation is the deviation from a literal rendering of *aśrutaṃ*, *amataṃ*, and *avijñātaṃ* in favour of a less literal but more accurate evocation. Müller ends his somewhat bitter controversy with his friend (Goreh) with the following notes:

> Dr Ballantyne therefore felt exactly what I felt, that in our passage a strictly literal translation would be wrong, would convey no meaning, or a wrong

[57]*The Upaniṣads*, trans. by Max Müller, rep. Dover Publications, 1962, v. 2., p. xii.

[58]Ibid. p. xiv.

[59]'Hast thou asked (of thy teacher) for that instruction by which what is not heard becomes heard, what is not comprehended becomes comprehended, what is not known becomes known' (*The Upaniṣads*, trans. by Max Müller, rep. Dover Publications, 1962, v. 2. p. xiv).

[60]'Have you enquired of your tutor about that subject which makes the unheard-of heard, the unconsidered considered, and the unsettled settled?' Müller notes that 'he (R. Mitra) evidently knew that Brahman was intended, but his rendering of the three verbs is not exact'. (*The Upaniṣads*, trans. by Max Müller, rep. Dover Publications, 1962, v. 2., p. xvi.).

[61]It is essentially identical to Goreh's.

[62]'Thou, O disciple, hast asked for that instruction whereby the unheard-of becomes heard, the inconceivable becomes conceived, and the unknowable becomes thoroughly known.' (*The Upaniṣads*, trans. by Max Müller, rep. Dover Publications, 1962, v. 2., p. xvi–ii).

meaning; and Mr. Nehemiah Goreh will see that he ought not to express blame, without trying to find out whether those whom he blames for want of exactness, were not in reality more scrupulously exact in their translation than he has proved himself to be.[63]

Müller's translation of Uddālaka's question sounds reasonable and even correct. However, his rendering of the meaning of Uddālaka's question within the narrative supports his assertion that '[s]urely Mr. Nehemiah Goreh knew that the instruction which the father refers to, is the instruction regarding Brahman, and that in all which follows the father tries to lead his son by slow degrees to a knowledge of Brahman.'[64]

Engaged in debate with his friend Goreh ('Now, before finding fault, why did he not ask himself what possible reason I could have had for deviating from the original, and for translating *avijñāta* by unknowable or what cannot be known, rather than by unknown, as every one would be inclined to translate these words at first sight?'), M. Müller defends his reading of *Chāndogya* 6.1.3. by imagining Śvetaketu's state of mind at the time he was being questioned by his father:

What could Śvetaketu have answered, if his father had asked him, whether he had not asked for that instruction by which what is not heard becomes heard, what is not comprehended becomes comprehended, what is not known becomes known? He would have answered, 'Yes, I have asked for it; and from the first day on which I learnt the Siksha, the A B C, I have every day heard something which I had not heard before, I had comprehended something which I had not comprehended before, I have known something which I had not known before.' Then why does he say in reply, 'What is that instruction?'

Müller's point is truly important; for indeed, he argues that the question is paradoxical for ordinary minds, and since the answer is brahman Śvetaketu cannot answer it. His entire argument is simple enough, since Uddālaka means the teaching of Brahman which makes everything known. However bold M. Müller's cognitive identification with Śvetaketu, his rendering of the narrative is indirect and follows too closely the transmitted (rather than the full) narrative. Müller strictly focuses on the intellectual content of the question, and does not attend to the context and meaning of Uddālaka's question in the narrative at this point. He heeds neither Śvetaketu's conceit, nor the implied influence of his surrogate

[63]*The Upaniṣads*, trans. by Max Müller, rep. Dover Publications, 1962, v. 2., p. xvii.
[64]Ibid., v. 2., p. xv.

father (his guru). The suggestive tension and aggression and general narrative significance of Uddālaka's question to his son elude Müller's attention altogether.

Reflecting on the fateful contest between father and son in another context, D. Shulman develops a reading-orientation which integrates narrative significance with 'philosophical contents' at the highest degree. Shulman succinctly describes the encounter between father and son as a 'cunning and potentially deadly linguistic duel. ...' This metaphor, expressive of Uddālaka's aggression of and attack on Śvetaketu's adoption of his surrogate father, is pertinent to the encounter between Uddālaka and Śvetaketu. Seeking efficient contextual reading of one of the most telling episodes in the Mahābhārata (the Yakṣa's questions to Yudhiṣṭhira), Shulman characterizes a challenging question (*praśna*) as follows: '*praśna* points to a baffling, ultimately insoluble crystallization of conflict articulated along opposing lines of interpretation.'[65] And finally, he characterizes the questioning event in the following integrative mood:

> Both questions and answers tend to the metaphysical, with the latent centre of meaning—the ultimate reality that is the true object of the quest—usually present only as a suggested power situated somewhere between the two explicit poles of the contest.[66]

Following such an integrative approach, I venture to end this chapter by telling my version of the full story of Uddālaka Āruṇi and his son Śvetaketu:

Śvetaketu was the son of the great and renown sage, Uddālaka Āruṇi. Though of such distinguished lineage and name, he did not undergo the *upanayana* rite of passage into the study of the Veda until he was about twelve years old. Rejecting him, his father did not initiate him and, consequently, did not instruct him either (though he was probably one of the most qualified gurus). Other adolescents of his age had already mastered one or two of the four Vedas, and had also left home. Revolting against his most successful, wise, and famous (domineering?) father, Śvetaketu is adamant in his refusal to leave home and study the Veda at the home of a professional teacher (guru). Feeling rejected by his father (who does not want to educate him), he resists going away as much as he can. Apparently a very intelligent youth, Śvetaketu might have used his father's judgement of the old Vedas (in their inferiority to the Upanishadic knowledge) as a

[65]'The Yakṣa's Questions,' in G. Hasan-Rokem and D. Shulman, eds, *Untying the Knot*, Oxford University Press, 1996, p. 153.

[66]Ibid., p. 153.

justification of his resistance to leaving home for *brahmacarya*. His father, disappointed and deeply concerned over his son's delinquent lifestyle, accuses him of latent *śūdra*hood. Unable to express his anger, denying these feelings, he says in an objective tone: 'In our family there are no fallen—unqualified, defective, hypocrite—brahmins who do not study, who are actually *śūdras*. Since you have not studied the Veda, and since in my family there are no members who are *brahmabandhu*, and since you are bound to become such a false brahmin, you are not my real son anymore. Go away and study! Then you may come home.' Deeply hurt, exiled from home, Śvetaketu goes away for twelve years. (Note that he is represented as having lived an equal number of years—twelve—with each of his 'fathers'.) He comes back knowledgeable, renewed with self-esteem, his teachers' support, and probably also revengeful. He does succeed in his study, and feels superior to his father, to whom he might have said: Indeed, I am not your real son now; for the teacher who has instructed me is my real father. On the road to reclaiming his son, Uddālaka inaugurates a Vedic contest and asks: 'But did you enquire about the teaching by virtue of which the unheard becomes heard, the unknown becomes known?' Conceding defeat, Śvetaketu says: 'Tell me that.' Śvetaketu becomes thus endowed with a new sense of inferiority, perceiving some relevance, meaning, and excellence in his father's questions. Concomitantly with his acquisition of newly sensed weakness (and therefore inferiority), he is open to the horizon of knowledge and excellence. Only after recciving a powerful answer does Śvetaketu renounce his teachers: 'They probably did not know that; for if they did they would have told me.' Thus, Śvetaketu returns to his father and asks to become his son again. This time, unlike the situation in Śvetaketu's adolescence, Uddālaka accepts him as son and pupil. Indeed, father and son stay together for many days; apparently they experience and experiment together for several months. They reflect on sleep, death and consciousness; they watch bees and honey, fruits and flowers. Śvetaketu tastes salt, fasts for two weeks by his father's side and recites the Veda. Closer and closer to his father, he becomes ripe, as it were, for leaving his inferiority behind and sharing knowledge and the mystery of his unity with his father. Thus Uddālaka shares being with his son, and says: 'That being, which is me, is also you, Śvetaketu' (*Tat tvam asi śvetaketo*).

Epilogue: Storytelling and Fearful Self-Understanding

The ancient Upanishadic rishis articulate an allegedly liberating truth (or meta-truth): You are what you know; or, you are what you understand; or, you become what you think you are. At moments, from our location on the margin, we think we understand such meta-statements and, to an extent, we do understand. In addition to psychological insights on the potency of self-image, etc., learning and education do change one's being. However, an otherness pervades, in the same way that one does not become an animal or plant in the process of learning about animals or plants. But the Upaniṣads ceaselessly repeat formulas such as: 'Whoever knows the oldest and the best becomes oldest and the best' (*yo ha vai jyeṣṭhaṃ ca śreṣṭaṃ ca veda jyeṣṭhaśca ha vai śreṣṭhaśca bhavati*).[1] 'Whatever victory belongs to brahman, and whatever he possesses, this victory and this possession he wins and obtains; who knows this (*ya evaṃ veda*), who knows this.'[2] Such assertions are invaluable references to the nature of Upanishadic knowledge and its experience by the Upanishadic sages.

While knowledge is potentially transformative and empowering, consciousness of self (*ātman*) is indeed revolutionary. Cessation of suffering, a sense of freedom, immortality, unsurpassed happiness, and glowing faces are among the expressions of this consciousness of *ātman*.

The Vedic culture of connections, resemblances and identities reaches a moment of breathtaking recognition of self and its identity with the

[1]ChU 5.1.

[2]KauU 1.7 (*sā yā brahmaṇo jitir yā vyaṣṭis tāṃ jitiṃ jayati tāṃ vyaṣṭiṃ vyaśnute ya evaṃ veda, ya evaṃ veda*).

unchanging source of everything, brahman.[3] This is the highest type of knowledge distinguished since the early Upaniṣads as different in quality from others. Its conceptualization by ancient Indian seers is different from our own associations of 'knowledge'. And yet, with respect to ourselves, statements such as the above ('you become what you think you are') are a viable starting point for scholarship and focused considerations and investigations pertaining to the nature of the consequential 'knowledge' or 'consciousness' (*jñāna, vijñāna, khyāti, vidyā, cit*) extolled in the Upaniṣads. Under-reading the Upanishadic stories is a salient source of the relative lack of intense reflection on the particular flavours of Upanishadic 'knowledge' (and especially 'knowledge' of self). However, 'knowledge of self' is indeed particularly difficult to conceptualize.

This is somewhat paradoxical, since what knowledge is seemingly more accessible than knowledge of self? One does not have to go far to gather much information. The self is common to all, and near, very near, singularly near; indeed, nearest to 'us'. And let's suppose also that it is difficult to think about this self, which is so elusive, without assistance. We look to the Upaniṣads for help, since, of course, the Upaniṣads point to this self. Readers of the Upaniṣads—margin-dwellers, as I have been calling them—know that their selves are present close to them. But alas! Our faces resist glowing; we read and reread the Upaniṣads and do not become immortal. This situation will not change for us margin-dwellers soon.

And yet we can attend more carefully to Upanishadic testimony about the nature of knowing the self. Obviously, though a Yājñavalkya, Satyakāma, Haridrumata, or Uddālaka would not recognize us as knowers of brahman (nay, they might not even accept us as disciples), they do speak with a perceptible intentionality. Moreover, powerful references and cues are included in their speech, attesting to the nature of Upanishadic 'experience of knowledge'. The most powerful among these references and cues are the Upanishadic stories. Indeed, these stories point to the 'conditions of knowledge', and the obscure, fateful and fearful nature of the encounter with one's self. Thus the Upanishadic narrative testifies to essential aspects of 'knowledge of self', the goal and achievement of the Upanishadic sages. The basic unit is, however, neither the 'story' nor the 'teaching'. It is both circumstances (narrative) of knowledge transmission

[3]These connections, resemblances, etc. 'collapse', as B.K. Smith says, into the single monistic identification of *ātman* with brahman. Cf. *Reflections on Resemblance, Ritual, and Religion*, Oxford University Press, 1989, p. 31.

and the 'contents' of knowledge. Closer reading of the Upanishadic testimony, which includes both the teaching and the narrative, is conducive to understanding the Upanishadic experience and knowledge.

The theme of inferiority is central to my understanding of the Upanishadic narrative. The Upanishadic heroes express the need to move from inferiority to excellence in their life stories, a characteristic common to all the dramatic situations in the early Upaniṣads. Research into the nature and meaning of these Upanishadic inferiorities—for there are several—helps illuminate the nature of 'knowledge' instrumental in their transcendence. Awareness of a pure self—as horizon, a standard in emergence, a teleology—combines with essential weaknesses to create and constitute the Upanishadic inferiority. The crux of the Upanishadic inferiority is the refusal to accept weaknesses such as death, binding desires, vanity, doubt, and even bodily weaknesses. In contrast to basic orientations of our own culture, the Upanishadic protagonists seem to pursue radical transformation and transcendence, resisting, for example, embodiment and its inferiorities. The articulation of this point in *Chāndogya* 8.12 is unambiguous; Prajāpati offers Indra a vision of playful existence, bursting with incredible freedoms:

> This body, Maghavan, is mortal; it is in the grip of death. So, it is the abode of this immortal and non-bodily self. One who has a body is in the grip of joy and sorrow, and there is no freedom from joy and sorrow for one who has a body. Joy and sorrow, however, do not affect one who has no body. The wind is without a body, and so are the rain-cloud, lightening, and thunder. These are without bodies. Now, as these, after they rise up from the space up above and reach the highest light, emerge in their own true appearance, in the very same way, this deeply serene one, after he rises up from this body and reaches the highest light emerges in his true appearance. He is the highest person. He roams about there, laughing, playing, and enjoying himself with women, carriages, or relatives without remembering the appendage that is this body.[4]

Upanishadic knowledge is renunciation of inferiority along with its vessel; it is transcendence of the very condition of inferiority. As a measure of the discrepancy between prevailing existence and heavenly otherness, this concept of inferiority is universally helpful for man's self-understanding, while not excluding the potential of individual standards and idiosyncratic experiences for breeding new modes of inferiority.

Recognizing the problematic of Müller's sometime-brahmin reader while embracing its spirit, this book has essayed to discuss and display a

[4]*The Early Upaniṣads*, trans. by P. Olivelle, Oxford University Press, 1998, p. 285.

reading approach which invites fuller understandings than those offered in the transmitted texts; these texts are reduced remnants, and reach most readers today as such. This return to the text, to narrative attention and close reading, offers a reading approach more accessible to margin-dwellers than Müller's Brahminic aspirations, while sharing his belief in the necessity of an empathic reading mode which can facilitate fuller understanding.

The Upanishadic sages spoke of self and knowledge. Sometimes, our rendering of their speeches looks too familiar and commonplace to be true, thus we know we are wrong and in need of a more attentive listening of the Upanishadic voices expressive of the sages' intentions. Their intentions—for they did have intentions—are precious yet barely accessible. Grounded in ancient tradition, the partial rejection thereof, revolt, crisis, discipline, and unusual experiences—the testimony of these voices is not easily retrieved and digested. Indeed, this inaccessibility of one's self is perhaps the most salient subtext of the Upaniṣads. While talking of the self, the Upanishadic story also enacts its inaccessibility.

Apart from intentions which naturally vitalize Upanishadic speech, the sages of old also had their characters, individuality and particular modes of inferiority. These aspects—also instructive and most closely connected with storytelling—are consistently and surprisingly under-attended in the exegetic tradition.

Let us briefly note an episode expressive of the precious value of narrative to the explication of the Upanishadic perception (or conceptualization) of knowledge. In this episode—as I set its boundaries here—the contents of knowledge (such as *Tat tvam asi*) are not discussed at all. It serves here as an example of a Upanishadic speech-act bearing exclusively on the nature of knowledge by the sheer force of the story.

In *Chāndogya Upaniṣad* 5.11.4 there is the story of Uddālaka Āruṇi who is reluctant to assume the function and role of transmitting knowledge to five great householders and scholars of the Veda who come in for instruction on the *ātman*. These five householders are mentioned by name;[5] they are not ordinary people. They have come to Uddālaka after considering the matter among themselves; What is our self (*ko na ātmā*)? What is brahman (*kiṃ brahmeti*)? Obviously, the choice of teacher is fateful. The five great householders have heard about Uddālaka as a sage devoted to study of the *ātman*, and have chosen him. On his part, Uddālaka is filled

[5]Prācīnaśāla Aupamanavya, etc. These names, however, are not mentioned elsewhere in the Upaniṣads.

with anxiety; 'These great householders and scholars of the Veda (*śrotriya*) will ask me questions. I may not be able to answer all of them' (*prakṣyanti māmime mahāśāla mahāśrotriyāḥ. Tebhyo na sarvamiva pratipatsye*). He decides to refer them to another teacher, the king Aśvapati Kaikeya; 'Ah! I shall direct them to another!' (*hantaham anyam abhyanuśāsānīti*).

This little anecdote provides a glimpse into the Upanishadic circumstances and perception of 'knowledge'. Whence Uddālaka's anxiety or sense of inferiority? What is the nature of this anxiety? These questions arise even from a reading of the transmitted story; the fuller one is almost beyond reach. Above all, we have no idea what Uddālaka's reference to another teacher means; does it signify conceding defeat?

The famous third section of the *Bṛhadāraṇyaka Upaniṣad* describes a contest among Vedic sages. In this fateful contest, Yājñavalkya is the big winner. In that context, Uddālaka—though he has witnessed Yājñavalkya's greatness—refuses to accept the latter's superiority, and he bitterly fights Yājñavalkya to the end (Uddālaka's silence).[6] He curses him, even tries to kill him.[7] Thus this testimony in BU 3 about the significance of not knowing, and particularly the significance of the fact of not knowing the self among other sages (with probably different ideas about the self), points to the Upanishadic sage's sensitivities, and in particular to Uddālaka's self-esteem, hatred of pretension and possibly—in the terms we have been using—a latent sense of inferiority. Thus, the pernicious insecurity about knowledge of self, culminating here in the refusal to teach and the surrender of authority to another sage, is not necessarily a trifle; it may signify an awesome occurrence and crisis in the life of a Upanishadic sage such as Uddālaka. (It is noteworthy that Uddālaka's insecurity is told in the *Chāndogya*, Uddālaka's 'home-Upaniṣad').

On the margin, the sense of wonder at the nature of knowledge does not arise in this case. One can hardly share the great teacher's pain and insecurity. In general, Upanishadic inferiorities—which determine and constitute the nature of Upanishadic 'knowledge'—are difficult to gauge, identify and engage with, and understand. Given the disconnectedness of context and 'knowledge' in the view from the margin, the circumstances in which knowledge emerges become mere curiosities. Referring to the story of Uddālaka's withdrawal from teaching as 'one of the simple scenes'

[6]See BU 3.7.

[7]See my 'The Upanishadic Story and the Hidden *Vidyā*; Personality and Possession in the *Bṛhadāraṇyaka Upaniṣad*,' *Journal of Indian Philosophy* 26, pp. 373–85.

(other 'small scenes' are stories such as those of Yājñavalkya and Maitreyī, Uddālaka and Śvetaketu, Satyakāma and his mother, etc.), Oldenberg tells the story thus:

Several intelligent gentlemen come together, all 'rich in possessions and learnedness'. They go together to a reputed expert; but he is not sure of his knowledge and sends them further to another person, strangely enough to a king. They find their goal with his help.[8]

We see here that Uddālaka's fear of not knowing the answers to the gentlemen's questions about the self does not rouse Oldenberg's curiosity about Uddālaka's character, plight, and self-understanding. Similarly distant from the narrative source are Oldenberg's consistent references to Uddālaka not by name but rather as 'the father', 'the sage', 'the reputed sage', etc. Oldenberg's under-reading of the story generally and of Uddālaka's individuality specifically prevent him from seeing the invaluable testimony of *Chāndogya* 5.11.4 about the more obscure and probably dangerous aspects of Upanishadic 'knowledge'.

Interestingly, Oldenberg (for the purpose of this discussion, a faithful representative of the scholarly margin) describes another reputed sage in a somewhat different, and more fully characterized, manner. Thus, he announces the identity of the first among the Upanishadic sages: 'Perhaps the first one of the beginning of a development to such a peerlessness was Yājñavalkya, the absolute knower, the supreme.'[9]

As we watch, Oldenberg continues thinking, and incrementally changes his verdict of the supreme and peerless, proceeding towards a more balanced and even ambivalent judgment. To begin with, Oldenberg says, Yājñavalkya is one of the Upanishadic sages who were human beings with normal infirmities:[10]

These thinkers of the Upaniṣads are, on the whole, no ideal personalities. The human infirmities, with which the urge for knowledge and deliverance was surely mixed among many without any contradiction—here as elsewhere—are candidly, one can say, unknowingly, sketched. Even the tinge of dirt that is used to belong to Indian asceticism, is not forgotten.[11]

[8]H. Oldenberg, *The Doctrine of the Upaniṣads and the Early Buddhism*, trans. by S.B. Shrotri, Motilal Banarsidass, 1991, p. 95.

[9]Ibid., p. 100.

[10]Oldenberg refers, no doubt, to BU 3.

[11]Ibid., p. 100.

Oldenberg's last note about the tinge of dirt of Indian mysticism is somewhat puzzling. What could he mean? He has in mind Yājñavalkya's personality, the supreme sage who is also a boastful man, humiliating his rivals:

A great role is played, coming down from the primordial period, by diligent passion for a sportive achievement of intellectual tournies, for a possible effective show that one knows what others do not know.[12]

After proceeding in his assessment of Yājñavalkya's personality according to the third section of the *Bṛhadāraṇyaka Upaniṣad*, Oldenberg reaches the apex of his narrative, ending his analysis of Yājñavalkya, the one who 'is a virtuouso, and wishes to be acknowledged as such' with a suggestion concerning what Yājñavalkya would have had passion for (had he not been restrained by the 'tinge of dirt that is used to belong to Indian asceticism'):

The vanquisher, however, now satisfied, takes over the honour beside the cows with gold hung on their horns. And should such wise men be unreceptive for such possession, the chance remains that he has more liking for a beautiful girl.[13]

Thus this sympathetic and most knowledgeable margin-dweller does not merely describe the sage's actual behaviour, but he also guesses what the great Yājñavalkya *would* do. Oldenberg does not hide his disappointment with the Upanishadic characters revealed in the Upanishadic storytelling: 'Thus indeed these narrations do not convince that the new ideals have taken possession of the whole man.'[14] However, Oldenberg—exceptional in his erudition and moral sensitivity—not only draws a distinction between Yājñavalkya's depraved personality and the rest of Upanishadic spirit, but seemingly forgives Yājñavalkya himself (apparently the 'old man [who] brings order to his house and wanders forth homeless from there'):

But then again: how genuinely, the desire for knowledge of great mysteries: how with convincing power it is expressed in many a passage! How the dialogues between father and son, husband and wife, the wise and his disciple prince are so distinct from the shallowness of those battles of words![15] The face of the one who understands Brahman glows. The old man brings order to his house and wanders forth homeless from there. What an awareness that here comes in question a possession which does not have its equal.[16]

[12]Ibid.
[13]Ibid.
[14]Ibid.
[15]Oldenberg refers here to the fight among the sages in BU 3.
[16]Ibid.

Complex indeed is Oldenberg's attitude; spiritual, enthusiastic, yet judgemental and condescending. In his treatment of the Upanishadic stories as sketches of small and simple scenes—irrelevant to understanding the contents and nature of knowledge—Oldenberg, like the mass of margin-dwellers, overlooks the significance of the Upanishadic storytelling in attempting to excavate Upanishadic experience and perception of knowledge. Due perhaps to a preference for explanation over understanding, the Upanishadic stories have been consistently under-read, relegated to the irrelevance of discarded packaging. Yet these stories, reflecting distant yet possibly highly valuable experiences of sages in the wilderness, invite our creative attention. Perhaps the translation of weakness into inferiority and the consequent seeking of excellence and knowledge is a course of reflective development that can offer its own contemporary resonance.

Bibliography

Primary Sources

Bṛhadāraṇyaka Upaniṣad

Chāndogya Upaniṣad

Kaṭha Upaniṣad

Kauṣītaki Upaniṣad

Muṇḍaka Upaniṣad

Taittirīya Upaniṣad

The Early Upaniṣads, Annotated text and trans. by P. Olivelle, Oxford University Press, New York, 1998.

The Laws of Manu, trans. by W. Doniger and B.K. Smith, Penguin Books, London, 1991.

The Principal Upaniṣads, edited by S. Radhakrishnan, George Allen & Unwin Ltd., London, 1953.

Sannyāsa Upaniṣads, trans. with an introduction by P. Olivelle, Oxford University Press, New York, 1992.

Śatapathabrāhmaṇa, trans. by J. Eggeling, Motilal Banarsidass, New Delhi, 1978.

Ten Principal Upaniṣads with Śāṅkarabhāṣya, Motilal Banarsidass, New Delhi, 1964.

The Thirteen Principal Upaniṣads, trans. by R.E. Hume, Oxford University Press, 1921.

Upaniṣads, trans. by P. Olivelle, Oxford University Press, New York, 1996.

The Upaniṣads, trans. by Max Müller, rep. Dover Publications, New York, 1962.

The Vedānta-Sūtras with the Commentary by Śaṅkarācārya, trans. by G. Thibaut, Oxford University Press, 1904, rep. Motilal Banarsidass, New Delhi, 1994.

Secondary Sources

Deussen, P. *Sixty Upaniṣads of the Veda*, trans. by V.M. Bedekar and G.B. Palsule, Motilal Banarsidass, New Delhi, 1991.

Flood, G. *An Introduction to Hinduism*, Cambridge University Press, Cambridge, 1998.

Gombrich, R.F. *Theravāda Buddhism*, Routledge, London, 1988.

Gonda, J. *Change and Continuity*, Mouton & Co., The Hague, 1965.

Gough, A.E. *The Philosophy of the Upanishads and Ancient Indian Metaphysics*, Trubner & Co., Leiden, 1882.

Halbfass, W. *India and Europe*, SUNY, New York, 1988.

______ *Tradition and Reflection*, SUNY, New York, 1990.

Handelman, D. and Shulman D. *God Inside Out: Śiva's Game of Dice*, Oxford University Press, New York, 1997.

Hasan-Rokem, G. and Shulman D., eds, *Untying the Knot: On Riddles and Other Enigmatic Modes*, Oxford University Press, New York, 1996.

Heesterman, J.C. *The Broken World of Sacrifice*, University of Chicago Press, Chicago, 1993.

Hiriyanna, M. *The Essentials of Indian Philosophy*, George Allen & Unwin Ltd., London, 1985.

Hopkins, T.J. *The Hindu Religious Tradition*, Wadsworth Publishing Company, Belmont, 1971.

Jung, C.G. *Jung of the East*, J.J. Clarke, ed., Routledge, London, 1995.

Keith A.B. *The Religion and Philosophy of the Veda and Upanishads*, Harvard University Press, 1925, rep. Motilal Banarsidass, New Delhi, 1989.

Knipe, D. *In the Image of Fire*, Motilal Banarsidass, New Delhi, 1975.

Kocmarek, I. *Language and Release*, Motilal Banarsidass, New Delhi, 1985.

Lanman, C. *The Beginnings of Hindu Pantheism*, Charles W. Sever, Boston, 1890.

Lannoy, R. *The Speaking Tree*, Oxford University Press, New York, 1971.

Lingat, R. *The Classical Law of India*, University of California Press, Los Angeles, 1973.

Mehta, P.D. *Early Indian Religious Thought*, Luzac & Company, 1956.

Mohanty, J.N. *Reason and Tradition in Indian Thought*, Oxford University Press, New York, 1992.

______ *Essays in Indian Philosophy, Traditional and Modern*, Oxford University Press, New York, 1993.

Oldenberg H. *The Doctrine of the Upaniṣads and the Early Buddhism*, trans. by S.B. Shrotri, Motilal Banarsidass, New Delhi, 1991.

Ranade, R.D. *A Constructive Survey of Upanishadic Philosophy*, Oriental Book Agency, Poona, 1926.

Rao, K.L.S. *The Concept of Śraddhā*, Motilal Banarsidass, New Delhi, 1974.

Ricoeur, P. *The Conflict of Interpretations*, Northwestern University Press, Evanston, 1974.

Said, E.W. *Orientalism*, Vintage Books, New York, 1978.

Shulman, D. *The Hungry God*, University of Chicago Press, Chicago, 1993

Smith, B.K. *Reflections on Resemblance, Ritual, and Religion*, Oxford University Press, New York, 1989.

Zaehner, R.C. *Hinduism*, Oxford University Press, 1966.

Zimmer, H. *Philosophies of India*, Meridian Books, New York, 1957.

Index

Absolute (Brahman) 5, 7, 8, 31, 32, 41, 42, 50, 54, 55, 57, 59, 63–4, 66, 80, 81, 88, 101, 105, 107, 108, 109, 110, 113, 114, 116, 120, 121, 127, 130, 131, 133
accuracy 83, 87
adhareya 15
Advaita Vedānta 105, 118, 119
afflictions 38
Aham brahāsmi 9, 114, 119
āhavanīya 43
air (*vāyu*) 22, 23, 28
Ajātśatru 54
akṣara 125
ambiguity 53, 54, 64, 65, 76
ambivalence 76, 78
ānanda 5, 94, 109
Ānandagiri 35
Angiras 125
Anima 114
antiquity 111
anvāhāryapacana fire 42
anxiety 43, 44, 134
Āpastambasūtra 121
apauruseya (non-human) 9–10
Āranyakas 8
Arthabhaga 91
arthavāda 10*n*
asceticism 64, 135, 136
aspirations 83
assertiveness 36
Aśvapati Kaikeya 134
Aśvins 82
ātman (consciousness, inner self) 3, 7, 9, 19, 25, 26, 27, 29, 31, 32, 34, 59, 65, 72, 74, 75, 78, 79, 80, 97, 101, 104, 105, 110, 113, 114, 115, 116, 130, 133
ātmavidyā 31
autonomy 12
avidyā 107
avijnāta 127
awareness 15, 24, 26, 73, 80, 82, 132

Bādarāyana 19
Bālāki 53
benevolence 9
Bhagvadgītā 8, 81
Bible 35
bigamy 74
bliss 1
bodilessness, state of 10*n*
body 132
Brahma 12
Brahmabandhu 102, 103, 122, 123, 129
Brahman, *See* Absolute

Brāhmanas 8, 55, 56, 63, 112, 114
Brahmanical theology 18, 71
Brahmasūtrabhāṣya 8, 19, 33, 119
Brahmavādinī 72
brahmin 102, 121, 123, 129
breath 41
Bṛhadārṇyaka Upaniṣad 8*n*, 66, 67, 70, 71, 72, 73, 74, 75, 79, 91, 115, 125, 134, 136
brahmacarya 121, 123, 129
Buddha 61, 77, 107

caste differences/system 49, 107
celestal light (*jyotiḥ*) 2
Chāndogya Upaniṣad 1, 8*n*, 9, 15, 17, 22, 28, 32, 37, 38, 45, 47, 50, 91, 132, 133, 134, 135
charity 78
Citra Gāṅgyāyani 91
cognition 28
collectedness 26, 27
commitment 6
communication 59
compassion 78, 106
compatibility (commensurability) of inferiority 77, 88
completion 2
conceptualization 105
confidence 2, 16
conflict 3, 20, 53, 69, 70, 74, 75, 76, 78, 79, 114, 118, 124, 128; and development, 37; personal and interpersonal, 8
connectedness 29, 57
consciousness (*jñāna*, *vijñāna*, *khyāti*, *vidyā*, *cit*) 23, 73, 76, 85, 105, 107, 108, 109, 129, 130, 131; after death, 67 *see also ātman*
context and metaphysics, connection 37
contextaulization, contextuality 35, 104
corruption 35
cosmos 120
crisis 71
cross-cultural contact 65

Dasgupta, S. 14
death 85, 87, 91, 92, 95, 132 *See also* Yama
'debt, theology of' 71
decontextualization 81, 118
depression 40
desires (*kāma*), 38–9, 40, 41, 42, 57, 88
destiny 119
detachment 108
Deussen P. 35, 60, 63, 66, 115
Dharmaśāstra 52
dialogue 3
dissatisfaction 78
divinity 116
duality, 76, 77, 118
dvija, *See* twice born

earth (*prthivī*), 42, *See also* five elements
Eckhart 61
ego, egoism 29, 60, 74, 106, 108
embodiment 132
emotions 83, 89
empathy 17
emptiness 16, 37
enlightenment (knowledge) 78, 113
entitlement, (*adhikāra*) question of 47, 49, 51, 54
equality 107; social and religious 20
excellence 2
excitement 29, 82
existence (*sat*) 5, 7, 11, 14, 22, 63, 86, 109, 115, 125, 132; painful 31; totality of 108
existential crisis 30
Existential Reading 11–3, 14, 16, 20
experience 12, 83, 117

faith *See śraddhā*

fasting 38, 40
finite being 26
fire (*agni/tejas*) 22, 28, 31, 42, 44, 45
five elements 22, 28
food (*anna*) 28, 31, 42, 114
form/content relationship 24
formalism 97, 98
freedom 13, 69–70, 109, 130, 132
frustration 73, 76
fulfillment 13, 21
fullness 26, 31, 35, 37, 38, 116

Gambhirananda, 35
Gārgī 58
gārhapatya 42
gati 43
generosity 17
guru/*śisya* communication 27

Halbfass, W. 86
happiness 125, 130
Haridrumata Gautama 46, 47, 51, 55, 131
harmony 73, 99
heaven, 43
hermeneutics 36, 81, 85, 86
hierarchy 21, 31, 34
Hindus, Hinduism 83, 89
Hiriyanna, M. 14, 59, 60, 64, 114
holism 16
hope (*āśā*) 31
householder (*gārhasthya*), 22, 23, 24, 70
human kind 6
human speech 104
Hume, R.E. 89
humility 17, 36, 58, 124

'I am Brahman' *See Aham brahāsmi*
identity 17, 20, 37, 42, 43, 63, 85, 108, 119, 130, 135
ignorance 7, 113
immortality 1, 17, 25, 29, 57, 58, 63, 66, 68, 70, 71, 73, 76, 78, 79, 80, 82, 91–2, 95–6, 103, 108, 130
impersonal traits 17
incomprehension 76
indestructibility 63
individualism, individuality 17, 36, 37, 42, 60, 73, 83, 87, 133
Indra 25, 34, 35, 132
Inferiority 2, 17, 21, 18, 22, 23, 26, 39, 45, 53, 54, 55, 57, 58, 62, 68, 80, 88, 97, 129, 132, 133, 134, 137; to excellence, 25
Infinity 109
infirmity 80
initiation *See upanayana*
inner conflict 26, 74
'inner crisis' 16
inner doubt 88
inner space 26, 27, 37, 41, 42, 57, 81, 83, 85, 89, 116
insecurity 43
insight 29
intelligence (*citta*) 31
intentionality 131
interconnectedness 83
interpretation 19, 21, 128

Jabālā 45, 46, 48–9, 50–51, 52
Jaivali Pravāhana 113
Janaka 53
Jānaśruti 1–3, 9, 15–20, 21, 22, 23, 24, 35, 53, 68, 118
Jesus 36
jñāna 10*n*, 14, 117 *See also* knowledge
joy 5, 101, 106, 132

kāma 40
Kant, Immanuel 7, 63, 77, 107
Karman, doctrine of 10*n*, 45, 65
Kaṭha Upaniṣad 80, 81, 88, 89, 92, 93, 96, 99
Kātyāyanī 66, 67, 72, 74, 78
Kaushītaki 114

Keith, A.B. 27, 35, 62–4, 65, 77–8
knowledge (*vidyā*, *jñāna*, *vijñāna*) 21, 88, 101, 105, 110, 117; context-free nature 4, 81; means of obtaining 17; power of 113, 116–7, 125, 130–7; practical dimension, 120; transmission 91, 131–2
Kṛta 2, 15, 18

language, action of 27, 31–2, 34, 37, 62, 65, 88
language-in-crisis 30
Lanman, C. 65
Lannoy, R. 83–5
learning 124–5, 130
liberation *See mokṣa*
lightning 43
lineage (*gotra*), 44, 45, 48, 82
loneliness 36
love (*kāma*) 75
lucidity 64

Mahābhārata 128
mahāvākya 9, 35, 119
Maitreyī 9, 15, 17, 24, 34, 36, 81, 83, 91, 96, 103, 104, 106, 118, 125,135
Māṇḍukya Upaniṣad 54
mantravid 34
mantravidyā 31
Manu 46*n*, 52, 69–70
Manusmṛti 40, 52, 69, 124
marginality, margin and centre, margin-dweller 30, 31, 32, 33, 34, 35, 83, 85, 111, 112, 113–4, 118, 131, 133, 136, 137
marriage 71
maya 65
meaning 4, 13, 26, 27, 32, 44, 53, 64, 68, 72, 83, 88, 106, 117, 119, 122, 123, 127, 128, 129, 132; and story, relationship 8, 34
meditation (*dhyāna*) 8, 22, 31
memory (*smara*) 31
'merger, theory of' (*saṃvargavidyā*) 22, 23
'message', 24, 60, 105, 118
'metaphysics', 1, 14–5, 23, 24, 25, 29, 32, 33, 34, 35, 36, 41, 50, 55, 60, 62, 77, 78, 81, 104, 105, 118
meta-testimony, 83
mind (*mānas*), 22, 31, 41, 78, 109; different states, 26, 40
mokṣa 5, 6, 10*n*, 29, 119
monism, 120
moon, 22, 23
morals, 74
mortality, 57, 80, 88, 96, 97
Moses 36
Müller, Max 35, 75, 89, 110–2, 113, 116, 121, 126–8; Brahmanic aspirations, 132, 133
multiplicity 102
Muṇḍaka Upaniṣad 125

Naciketas 24, 34, 57, 80, 88, 90–9, 103, 104, 113, 118
Nārada 15, 24, 30, 31, 34, 36, 40, 57, 58, 68
narrative 19, 20, 24, 25, 26, 27, 30, 88, 117, 118, 132, 133; significance of 17; complexity 37
Nazarene 65
neo-Hinduism 12
New Testament 35
non-being, state of 123
non-duality 78, 104

objective entity and subjective me 42
Oldenberg, H. 35, 36, 66, 74, 77, 101, 106, 135–7
Olivelle, P. 35, 74, 86, 88, 89, 99
Om 8*n*, 80
ontology 24
openness 5,11, 23, 24, 26, 39, 67, 80, 83
Orientalism 85

orientation 24
otherness, 7, 11, 62, 82, 83, 87, 101, 116, 117, 130, 132

patience 27
phenomenal world (*saṁsāra*), 5–6, 8; and *mokṣa*, 10*n*
phenomenology 117
Plato 7
Pleasure 29
power (*bala*) 1, 2, 4, 21, 31, 115
Prajāpati 132
prakṛta 14, 39
prāmāna 14
prāṇa 41, 43
praśna 128
Pratardana 9
pratijñā 45
pride 88
procreation 87
progeny 42
puruṣa, 42, 92

Radhakrishnan, S. 6, 14, 15, 35, 44, 72, 75, 89, 92, 93, 97–9, 107, 109
Raikva 2–3, 9, 15, 16, 17–20, 21, 22, 23, 24, 53, 118
Ranade, R.D. 35, 73, 74, 75
reality 16, 50, 57, 101, 104, 109, 115, 117, 128; non-dual 25; unity of 20
reasoning (*yukti*) 112, 119
rebirth, cycle of 87, 120
receptivity 12, 26
reciprocity 37
recollectedness 29
reconciliation 43, 107
refinement 24
relevance 104
renunciation 22, 23, 26, 27, 29, 35, 37, 41, 44, 76, 88, 132
Ṛg-Veda, 70, 89
ritualism 99, 122
Roy, Rammohan 89

sacrifice 87, 92, 97, 99, 122
Said, Edward 85
salvation 120
Samkhyas 10*n*
Sanatkumāra 24, 30, 31, 34
Śaṅkara 2, 8, 9, 11, 12, 15, 17, 18, 19, 20, 21, 24, 27, 28, 30, 33, 35, 39, 40, 43, 47, 48, 49, 50, 51, 52, 53, 94, 107, 108, 119, 123
Sat 28, 110, 112–3
Śatapathabrāhmaṇa 8*n*, 82
satisfactions 81
satya 5, 109, 115
Satyakāma Jābāla 15, 24, 45, 46, 58, 66, 68, 82, 85, 103, 118, 131, 135
Śaunaka 125
Sāvitrī 121
Sāyana 89
Schopenhauer 60, 61, 106
Secondary Reading, 6, 7, 14
self, Self, Inner self, selfhood, (*pratyagātman*) 1, 2, 9, 22, 26, 27, 29, 30, 31, 32, 23, 34, 36, 37, 41, 56, 67, 68, 75, 76, 77, 78, 80, 85, 86, 88, 104, 105, 106, 107, 112, 113, 115, 119, 120, 125, 130, 131, 132, 133, 134, 135; confidence, 2, 51; and cosmos, equation, 110; criticism, 112; forgetfulness, 5; fulfillment, 21; image, 130; identity, 6, 42;—with the Real, 8; realization, 75; reflectivity, 7; as subjectivity, 63–4; unknowability, 29, 30
self-esteem, 16, 18, 19, 134
self-understanding 1, 28, 30, 38, 83, 86, 111, 113, 117
sharing 106, 18
siddhi 95
social identity 48, 49
social organization 83
śoka (sorrow) 8, 31, 85, 132
soul 77, 112
space (*ākāśa*) 31, 41, 42, 43

speech 5, 15, 22, 23, 24, 76, 77
speech-act 29, 31, 32, 33, 37, 39, 45, 102, 104, 108, 133
Spinoza 7
spiritual satiety 35
spiritual truths 6
spirituality 58, 68, 73, 85
śraddhā 17, 89, 91, 92, 93, 94, 95, 96, 99, 103
śrutiliṅga 49
story, storytelling 15, 17, 23, 26, 27, 30, 32, 33, 36, 37, 74, 81, 83, 88, 92, 117, 124; abstract instruction 1; dissociation from message 21, 24; relevance to nature and power of Upanishadic knowledge 1–3, 7, 8; and teaching, imbalance 119
subjectivity 29, 63–4, 97
śūdra, *śūdra*hood 18, 19, 20, 37, 45, 46, 47, 48, 49, 50, 51, 122, 123, 129
suffering 15, 17, 73, 78, 87, 130
sukhāvabodha 17
sun (*āditya*) 22, 23, 42
svātantrya 10*n*, 21
Śvetaketu 10, 11, 17, 24, 27, 28, 32, 34, 35, 36, 57, 58, 70, 91, 101–3, 106, 107, 108, 113, 115, 118–29, 135
Swahananda, Swami 44
symbolism 8*n*
sympathy 6

Tagore, Debendranath (1817–1905) 11–3
Tagore, Rabindranath 6, 12
Taittirīya Brāhmaṇa 89
Taittirīya Upaniṣad 29, 108, 114
tapas 39
Tat tvam asi 9, 10, 27, 28, 31, 32, 33, 35, 60, 101, 102, 106, 107, 108, 110, 113, 114, 115, 116, 118, 119, 125, 129, 133
teacher 38–41, 44; and narrative 132
technology 132
testimony 83
textual intelligibility 25
textual interaction 4
'theological empiricism' 120
'This self is the Absolute' (*ayam ātma brahma*) 9
'Thou are That' *See Tat tvam asi*
totality 120
Traditional Reading 4–6, 7, 8, 9, 11, 12, 14, 17, 19
tranquility 27
transcendence 37, 54, 55, 132; of conceptualization 118
transcendental otherness 7
transformation, process 7, 42, 54, 60, 85, 132
translation, translator 87, 88
truth, truthfulness 23, 44, 45, 46, 47 50, 51, 52, 53, 62, 63, 65, 88, 91, 109, 117, 118, 130
twice-born (*dvija*) 60–70, 121

Uddālaka Āruṇi 9, 10, 11, 24, 27, 28, 32, 33, 34, 35, 36, 58, 60, 61, 91, 101–3, 106, 107, 108, 112, 114, 115, 118–29, 131, 133, 134, 135
ultimate reality 5, 75, 88
unborn (*aja*) 81
under-read, under-reading 101, 102, 103, 104, 106, 116, 130, 137
understanding 14, 26, 28–30, 31, 33, 41, 42, 45, 57, 67, 68, 73, 76, 82, 85, 86, 90, 103, 106, 119, 123; knowledge based 21
uniformity 88
union 108
unity 50, 76–7, 78, 88, 101, 107–8, 109, 112, 116, 118
unity-of-speech act (*ekavākyatā*), 19
universality 116
Universe 115, 119
unreality 97

Upakosala 57, 58, 66, 83, 103, 106, 118
upanayana 121–3
urbanization 87
Uśan Vājaśravas 88, 89, 92, 93, 95–7, 99
Uṣasta Cākrāyaṇa 29, 66

Vācaspati Miśra 35
Vājaśravas *See* Uśan Vājaśravas
'value free' 81, 85
values 70, 73, 74, 81, 83, 97, 110, 126; disintegration 57, 62;
vānaprastha 70
Varṇāśrama 45, 47, 49
Veda, Vedas, Vedic 8, 9, 11, 20, 28, 30, 40, 45, 69–70, 71, 102, 103, 114, 121, 122, 123, 125, 128, 133; culture and society 70, 73; intentions 5; multivocality 10; orientation 83; otherness 1, 4, 5, 14, 17; ritualism 110; speech 9–10
Vedānta 8, 10*n*, 12, 14, 104, 107, 120
vidyā 104, 107, 113, 117
Virocana 35, 36
vital breathing (*prāṇa*) 22, 23
vital force (*prāṇa*) 31
vocal (and personal) multiplicity 82
vulnerability 36

water (*āpah*) 22, 28, 31
wealth 104
Weber, Max 86
West: consciousness of India 36; value-free tradition 14
wholeness (*ekavākyatā*) 16*n*, 21
wilderness 28
will (*saṅkalpa*) 31, 35
wisdom 97, 101, 107, 119, 115, 120
wish (*icchā*) 39, 40
women, status of 72

Yājñavalkya 9, 24, 29, 30, 33, 36, 58, 61, 63, 65–79, 83, 91, 103, 104, 106–7, 118, 120, 125, 131, 134, 135, 136
Yama (King of Death) 24, 80–1, 89–91, 93, 94, 96, 97, 99, 113, 118
Yoga 108

Carleton College Library
One North College Street
Northfield, MN 55057-4097

WITHDRAWN